Smart and Advanced Developments in Textile Materials

Smart and Advanced Developments in Textile Materials

Edited by
S. Kannadhasan
R. Nagarajan
Jacob Abraham
Kanagaraj Venusamy
Bharathi M. L.

CWP

This edition has been published by Central West Publishing PTY LTD, (ABN 13 683 898 722) Australia
© 2026 Central West Publishing PTY LTD

For more information about the books published by Central West Publishing PTY LTD, please visit https://centralwestpublishing.com

Disclaimer
Every effort has been made by the publisher, editors and authors while preparing this book, however, no warranties are made regarding the accuracy and completeness of the content. The publisher, editors and authors disclaim without any limitation all warranties as well as any implied warranties about sales, along with fitness of the content for a particular purpose. Citation of any website and other information sources does not mean any endorsement from the publisher, editors and authors. For ascertaining the suitability of the contents contained herein for a particular lab or commercial use, consultation with the subject expert is needed. In addition, while using the information and methods contained herein, the practitioners and researchers need to be mindful for their own safety, along with the safety of others, including the professional parties and premises for whom they have professional responsibility. To the fullest extent of law, the publisher, editors and authors are not liable in all circumstances (special, incidental, and consequential) for any injury and/or damage to persons and property, along with any potential loss of profit and other commercial damages due to the use of any methods, products, guidelines, procedures contained in the material herein.

A catalogue record for this book is available from the National Library of Australia

NATIONAL
LIBRARY
OF AUSTRALIA

ISBN (print): 978-1-922617-75-0

Preface

The aim of this book is to spread knowledge and real-world impact in an atmosphere of true international cooperation between scientists, engineers and have interactive sessions by bringing together the researchers, international societies and industrial heads to discuss the latest developments and innovations. It is a great platform for companies as well as institutions to represent their research services, products, innovations and research results in the fields of Textile Engineering. The development of technical textiles industry is only possible in an environment that supports interdisciplinary networks between researchers, engineers, technicians and industries. This can be done by the re-engineering of a traditional textile company, by the integration of textile processes by a user company, for whom these new materials provide an improved solution compared to traditional materials. The issue is to facilitate interaction between researchers and professionals working in the field of textile engineering. The special issue will provide an international forum for professionals from academia, industry and governing bodies to discuss their problems and share achievements in order to develop the nations through recent manufacturing techniques and technologies. This book may be helpful for both students and academics due to the wide variety of subjects addressed.

Table of Contents

Chapter 1

The Programmable Product Sorter Line Following Robotic Arm

Abhilash V Pandiankal,
Mar Augusthinose College, Ramapuram, Kottayam Dist., Kerala, India

Abstract

The Programmable Product Sorter Line Following Robotic Arm is one of the innovative ideas by using latest sensors, processors, advanced mechanics and strong program. This concept is very useful in industrial areas and paves the way for saving time. Programmable Product Sorter Line Following Robotic Arm (PPS Line Following Robotic Arm) is a type of robot with "pick and place mechanism" designed to ease the displacement of objects from one position to another as desired, using line follower mechanism. The working model of this concept described in this chapter. This system is developed by using IR sensor, RF ID, Arduino and servo motor mechanism. Based on the needs we can use other sensors and more powerful motor mechanism; this concept can be made more effective and customized.

Introduction

Now days Robot place a vital role in all the fields. Robot is a machine resembling a human being and able to replicate certain human movements and functions automatically. Development Electronics technology now causes digital technologies to lead the world. Development of strong programming language combined with digital technologies leads Artificial Intelligence closer to human. This will cause development of AI based robotics systems, which are helpful in reducing the risk factors in sophisticated working environments. Robotics are widely used in industrial automation, Vehicle automation, medical fields, Navigational fields, and home automation. Artificial intelligence capabilities are developed by using microcontrollers, microprocessors, and digital signal processors. Due to the improvisation of nanotechnology leads speed of operation and leads to create better performance of AI.

A Robot is a virtually intelligent agent capable of carrying out tasks robotically with the help of some supervision. Robots can be classified as autonomous, semiautonomous and remotely controlled. Robots are widely used for variety of tasks such as service stations, cleaning drains, and in tasks that are considered too dangerous to be performed by humans. The Programmable Product Sorter Line Following Robotic Arm (PPS LINE FOLLOWING ROBOTIC ARM) is a type of robot with "pick and place mechanism" designed to ease the displacement of objects from one position to another as desired, using line follower mechanism. Concept of working of line follower is related to light. We use here the behaviour of light at black and white surface. When light fall on a white surface, it gets reflected and in case of black surface light is completely absorbed. This property is used in building a line follower robot. In order to sort the object and move from one place to another, we use an RFID system (Radio Frequency Identification). Each RFID tag has a unique code which is different from one another. Pick and Place robot is successfully used to move the specimen that has been mounted RFID tag, to a predetermined position according to the group of objects and whether the position is empty or not. Pick and place robots are commonly used in modern manufacturing environments. Pick and place automation speeds up the process of picking up parts or items and placing them in other locations. Automating this process helps to increase production rates. Pick and place robots handle repetitive tasks while freeing up human workers to focus on more complex work.

Methodology

The main part of the design of PPS LINE FOLLOWING ROBOTIC ARM is ATMEGA-328p micro-controller which coordinates and controls the product's action. This specific micro controller is used in various types of embedded applications. Robotics involves elements of mechanical and electrical engineering, as well as control theory, computing and now artificial intelligence. According to the Robot Institute of America, A robot is a reprogrammable, multifunctional manipulator designed to move materials, parts, tools or specialized devices through variable programmed motions for the performance of a variety of tasks. The robots interact with their environment, which is an important objective in the development of robots. This interaction is commonly established by means of some sort of arm and gripping device or end effectors.

The IR sensor and RF Id sensors will help to read data from the outside word. Both sensors will help to read and identify the objects. The robotic arm mechanism for pick and place is controlled by using servo motor.

The block diagram of Programmable Product Sorter Line Following Robotic Arm is shown in figure 1. This system consists of a CPU, Path controller mechanism. Vehicle control Mechanism, Arm Control Mechanism, Object recognition system and a power supply. Figure below represents a generalized block diagram representation of Programmable Product Sorter Line Following Robotic Arm.

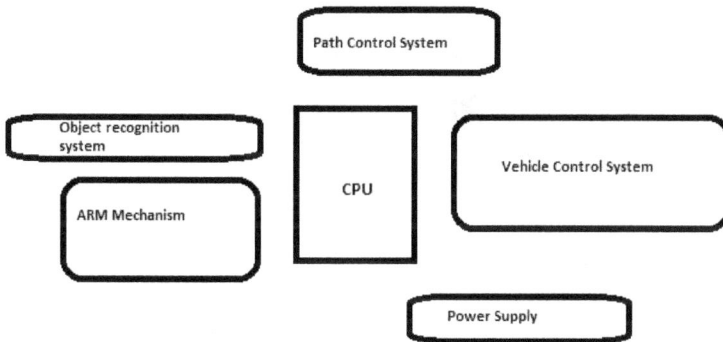

Figure 1. Block diagram.

The central processing unit of this system is ATMEGA 328 based Arduino microcontroller, which control all the operations performed by PPS LINE FOLLOWING ROBOTIC ARM. Path control system decides the vehicle movement. An IR sensor is used here to sense the object in its path. The object recognition system is designed by using RF ID Module. The movement and all functions of Arm mechanism is controlled by using servomotors. Dc motors with driver mechanism control the movement of the vehicle. A 12-volt power supply produces sufficient energy for the entire systems.

Functional Block Diagram

The functional block diagram of Programmable Product Sorter Line Following Robotic Arm is divided into six major blocks. The com-

3

munication between each block is carried out by appropriate data cables. Figure 2 shows the functional block diagram of Programmable Product Sorter Line Following Robotic Arm.

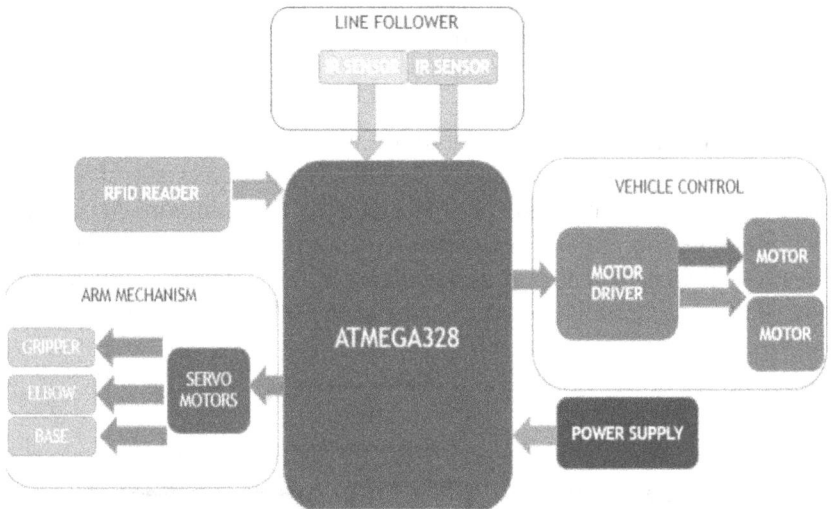

Figure 2. Functional block diagram.

Block Diagram Explanations

Arduino Uno R3 is used here as CPU. This is a versatile microcontroller and is equipped with both ATmega328P and the ATMega 16U2 Processor. This board gives you 14 digital input/output pins out of this can be used as PWM outputs, 6 analog inputs, a 16 MHz quartz crystal, a USB connection, a power jack, an ICSP header and a reset button. This board give all support as a microcontroller. Almost all functions are embedded within it. Its ease of use as microcontroller makes it too popular. It can be connected to computer through a USB Cable. This microcontroller consists of one built in USB B connector, 13 built in LED pin, Universal asynchronous receiver and transmitter (UART), accommodate synchronous serial data by usingSerial Peripheral Interface (SPI) protocol. Each I/O pin Carries 20 mA currents with an I/O voltage of 5V. The main processor ATmega 328p operating with a speed of 16 MHz and USB- Serial processor ATmega16U2 also operating with a speed of 16 MHz. ATmega16U2Microcontroller consists of an internal memory size of 2KB SRAM, 32KB FLASH and 1KB EEPROM. This will lead to the

4

processor to improve its action. Figure 3 shows the bus topology of Arduino Uno R3.

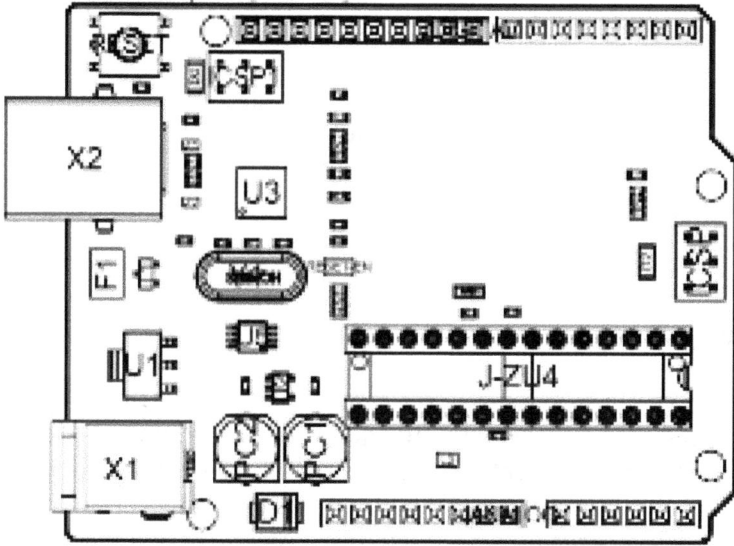

Figure 3. Arduino Uno R3 board topology.

Line Follower

An IR sensor is used here as an obstacle sensor of robotics path. An IR sensor is a device which emits light in order to sense some object of the surroundings. It can identify the heat of an object as well as it can detects the motion. All the objects have some radiation elements in the form of thermal radiation usually in the range of infra-red frequencies. The radiations which are in Infrared range is not visible to human eyes, but an Infrared sensor can detect this radiation and find out the obstacles in paths.

The IR sensor consists of two diodes one is used for emitting light, so it is an LED. The second diode is a photodiode and is sensitive to IR light of the same wavelength which is emitted by IR LED. Here the photo diode acts as a light detector. Its resistance and output voltages changes in proportion with the Infrared light which falls on the photodiode.

Working of IR Sensor is illustrated in Figure 4.

5

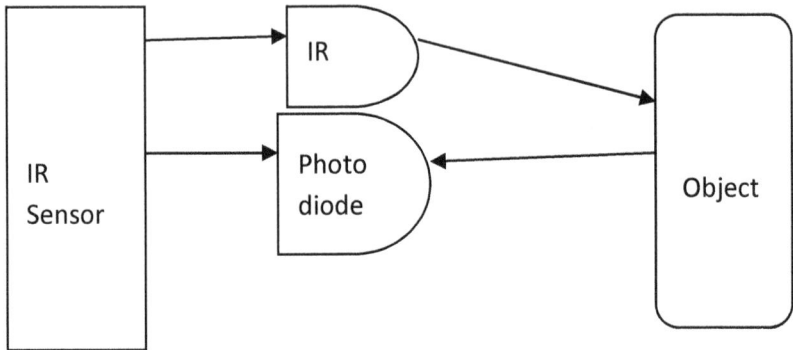

Figure 4. Working of IR sensor.

When the radiation from the IR transmitter reaches the object, some of this radiated light reflects back to the IR receiver. Based on the intensity of the received lights in the IR Receiver the output of the sensor changes its conditions.

An infra-red detection system consists of five elements which are:
- An infrared source
- A transmission medium
- An optical component
- Infrared detectors
- Signal processing

Coding of IR Sensor

```
void setup() {
  // put your setup code here, to run once:
pinMode(4,INPUT);
pinMode(12,OUTPUT);//LED
}
void loop() {
  // put your main code here, to run repeatedly:
if(digitalRead(4)==LOW){
digitalWrite(12,HIGH);
}
else{
digitalWrite(12,LOW);
```

}
}

RF ID Sensor

RFID is a short form of radio-frequency identification which is the wireless non-contact use of radio frequency waves to transfer data. Low Frequency, High Frequency, Ultra High Frequencies are the three frequency ranges used in RFID systems. Radio Frequency Identification system consists mainly two parts a transponder and a transceiver. Transponder, which is a tag attached to an object to be identified, and a Transceiver also known as interrogator or Reader. A Reader include Radio Frequency module and an antenna which is capable of generating high frequency electromagnetic field. Whereas the tag is usually a passive device, which does not contain a battery. It contains a microchip that stores and processes information, and an antenna to receive and transmit a signal. The information encoded in the tag can be read by placing it close to the reader. The Reader generates an electromagnetic field which in turns causes the electrons to move through the tag's antenna and subsequently power the chip.

The stored information from the powered chip inside the tag responds to the reader and sends the stored information. This information is sent to the microcontroller chip for further operations.

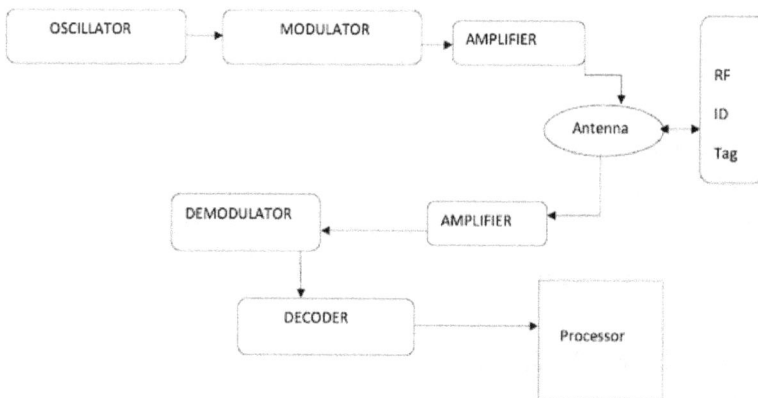

Figure 5. Block diagram of an RFID sensor.

Coding of RFID

```
#include <SPI.h>
#include <MFRC522.h>
#define SS_PIN 10
#define RST_PIN 9
MFRC522 mfrc522(SS_PIN, RST_PIN);   // Create MFRC522 instance.

void setup()
{
Serial.begin(9600);  // Initiate a serial communication
SPI.begin();     // Initiate  SPI bus
  mfrc522.PCD_Init();   // Initiate MFRC522
Serial.println("Approximate your card to the reader...");
Serial.println();

}
void loop()
{
  // Look for new cards
if ( ! mfrc522.PICC_IsNewCardPresent())
  {
return;
  }
  // Select one of the cards
if ( ! mfrc522.PICC_ReadCardSerial())
  {
return;
  }
  //Show UID on serial monitor
Serial.print("UID tag :");
  String content= "";
byte letter;
for (byte i = 0; i< mfrc522.uid.size; i++)
  {
Serial.print(mfrc522.uid.uidByte[i] < 0x10 ? " 0" : " ");
Serial.print(mfrc522.uid.uidByte[i], HEX);
content.concat(String(mfrc522.uid.uidByte[i] < 0x10 ? " 0" : " "));
content.concat(String(mfrc522.uid.uidByte[i], HEX));
```

```
}
Serial.println();
Serial.print("Message : ");
content.toUpperCase();
if (content.substring(1) == "BD 31 15 2B") //change here the
UID of the card/cards that you want to give access
  {
Serial.println("Authorized access");
Serial.println();
delay(3000);
  }

else {
Serial.println(" Access denied");
delay(3000);
  }
}
```

Vehicle Control

The speed control of motor is achieved by using DC Motor with a gear system. We can control the speed of the DC motor by simply controlling the input voltage to the motor. The input voltage control is achieved by using PWM (Pulse Width Modulation) technology. Here we use 12-volt Dc motor system. The pulse width modulation is a technique which allows us to adjust the average value of the voltage by turning on and off the power at a faster rate. The average voltage depends on the duty cycle. The PWM output from Arduino is connected to the base of transistor or the gate of a MOSFET. And the speed of the motor is controlled by controlling the PWM output. In order to work properly Arduino GND and the motor power supply GND should be connected together. For the speed control and direction control here use a driver circuit (L298N Driver). The dual H-Bridge of L298N Driver allows speed and direction control of two DC motors at the same time. So, this driver system acts as a gear mechanism for vehicle.

Arm Mechanism

The Arm mechanism is controlled by using servo motors. A servo motor is made by using DC motor which is controlled by a variable resistor and some gears. It works on the principle of Pulse Width Modulation. That is its angle of rotation is controlled by the duration of applied pulses. Servo motors specially designed to use in motion control applications. So, it requires high accuracy positioning, quick reversing and exceptional performance.

Coding For Servo Control

```
#include <Servo.h>
  Servo myservo; // create servo object to control a servo
intpotpin = 0; // analog pin used to connect the potentiometer
intval; // variable to read the value from the analog pin
void setup() {
myservo.attach(9); // attaches the servo on pin 9 to the servo object
}
void loop() {
val = analogRead(potpin);
  // reads the value of the potentiometer (value between 0 and 1023)
val = map(val, 0, 1023, 0, 180);
  // scale it to use it with the servo (value between 0 and 180)
myservo.write(val); // sets the servo position according to the scaled value
delay(15);
}
```

Working of PPS Line Following Robotics Arm

The main brain of this project is the Arduino, PPS LINE FOLLOWER ROBOT use pick and place mechanism for displacement of an objects from one position to another. Concept of working of line follower is related to light. We use here the behaviour of light at black and white surface. When light fall on a white surface it gets reflected and in case of black surface light is completely absorbed. This property is used inbuilding a line follower robot. Then we use an Op-Amp to check for change in voltage across the IR Receiver, so that if a fire is detected the output pin (DO) will give 0V(LOW) and if the is no fire the output pin will be 5V(HIGH).

In order to sort the object and move from one place to another, we use an RFID system (Radio Frequency Identification). Each RFID tag has a unique code which is different from one another. Pick and Place robot is successfully used to move the specimen that has been mounted RFID tag, to a predetermined position according to the group of objects and whether the position is empty or not.

So, in this project the motion of the robot is thus controlled by the atmega32 microcontroller, and the working is made possible through the help of IR sensor. The application of the PPS LINE FOL-LOWER ROBOT can be easily altered by make changes in the program which is burned into the microcontroller.

Flow Chart

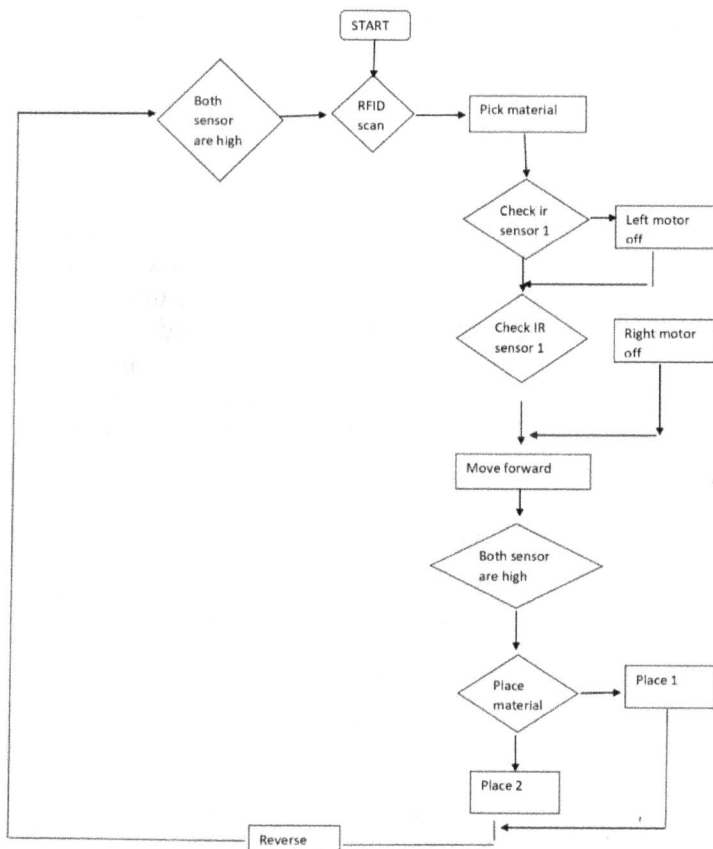

Result

After perfect coding the designed system is working properly. We use 12v power supply system for this designing. The arm mechanism is capable of handling objects with 1 kg and its movable all directions as per the instructions given from the CPU. As per the experiences from this designing work we can design its by our own needs by using appropriate sensors and motor Mechanism. Like every newly designed system this system also has some merits and demerits. The process speed is increased with the use of high-speed embedded system. Also, it can able to work in hazardous situations. High productivity of this system will make itself as very cost effective. Skill sets of employee, latest safety methods, budgeting; work-flow managements are some of the challenges faced during the implementation of robotic automation system. It causes potential job losses.

Conclusion

This project presents a pick and place robot using IR sensor and it is designed and implemented using arduinounoR3. The IR sensor has an IR Receiver (Photodiode) which is used to detect the reflected signal. In order to sort the object and move from one place to another, we use an RFID system (Radio Frequency Identification). Each RFID tag has a unique code which is different from one another. Pick and Place robot is successfully used to move the specimen that has been mounted RFID tag, to a predetermined position according to the group of objects and whether the position is empty or not.

Experimental work has been carried out carefully. The result shows that higher efficiency is indeed achieved using the embedded system. The proposed method is verified to be highly beneficial for security purpose and industrial purpose.

References

1. https://www.microchip.com/en-us/product/ATmega328P
2. University Malaysia Perlis,UNIMAP, "RobotCompetition,Theme& Rules", 2009.
3. Veselý, "Implementation of Micromouse Class Robot".

4. William Dubel, Hector Gongora, Kevin Bechtold and Daisy Diaz, "An Autonomous Robot".
5. John Iovine, "PIC Robotics: A Beginner's Guide to Robotics Projects Using the ARDUINO", McGraw Hill, 2004.
6. EaAi Choon, "Dc Motor Speed Control Using Microcontroller ", UniversitiTeknologi Malaysia, 2005.
7. Proteus PCBDesignPackages,"http://www.labcenter.co.uk/products/vsm_overview.cfm"
8. Custom Computer Cervices, "http://www.ccsinfo.com".
9. UIC00AUSB ICSP PIC Programmer,"http://www.cytron.com.my/usr_attachment/UIC00A_&_UIC-S_User_Manual.pdf
10. http://www.hobbyengineering.com/rmapIndex.html
11. www.wikipedia.com, robot, its benefits and its recent developments.
12. "MECHATRONICS" ALPHA I
13. http://www.robotics.com/robomenu/index.html
14. http://www.arrickrobotics.com/robomenu/index.html
15. Contest, http://www.trincoll.edu/~robo

Chapter 2

Leap Motion Sensor and Rpi based Smart Writing Technology using Convolutional Neural Network

Vijo M Joy[1], Joseph John[2]
Department of Electronics, Aquinas College Edacochin, Cochin-10, India
[2]Post Graduate Department of Physics, Aquinas College Edacochin, Cochin -10, India

Abstract

Smart Writing creates a virtual touch surface in the air. By touching this space, we can create virtual touch events. This can be done by using a Leap Motion Sensor and a Raspberry Pi. The Leap Motion Sensor detects the motion of out hand/pen and transfers the signals to the Raspberry Pi. From there the corresponding image will be created and displayed on the screen. Henceforth leads to the identification of the character. Mixed Reality (MR) opens a new dimension for Human Computer Interaction (HCI). Combined with computer vision (CV) techniques, it is possible to create advanced input devices. This project describes a novel form of HCI for the MR environment that combines CV with MR to allow a MR user to interact with a floating virtual touch screen using their bare hands. The system allows the visualization of the virtual interfaces and touch screen through a Head Mounted Display (HMD). Visual tracking and interpretation of the user's hand and finger motion allows the detection of key presses on the virtual touch screen. Artificial Intelligence is used to train the system and to improve the accuracy.

Introduction

Computer vision techniques are the least intrusive for interaction in a MR environment. In the case of hand-based interaction, some require the use of special gloves or markers attached to the hand to facilitate the interaction. This is less intrusive than the use of data gloves or 3D mice, as these special gloves or markers are lighter and do not need any wiring. However unadorned hand tracking can make it possible for a MR user to interact with the virtual and the real in the same way, by using their bare hands. Gesturing is a natural part of human communication and becomes more and more important in AR/VR interaction. The Leap Motion Controller is a new device developed for gesture interaction by Leap Motion. The device has a small dimension of 0.5x1.2x3 inches. To use the Leap Motion Controller, the user needs to connect it to a computer by USB. Then the users put hands on top of the Leap Motion Controller. Figure 1 gives the block diagram of the Leap Motion Controller [1-4].

The Leap Motion Controller could detect palm and fingers movements on top of it. The tracking data which contains the palm and fingers' position, direction, velocity could be accessed using its SDK. According to the Leap Motion Controller provides a detection accuracy of about 200m. The newest version of Leap Motion Controller Orion currently does not provide gesture recognition. We are trying to implement the gesture recognition by ourselves. Figure 2 gives the Leap Motion Coordinate system [5][6].

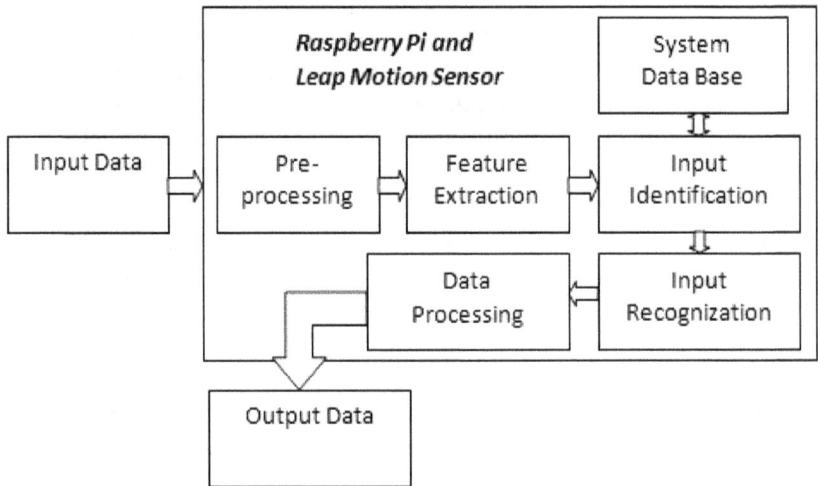

Figure 1. Block diagram of Leap Motion Controller.

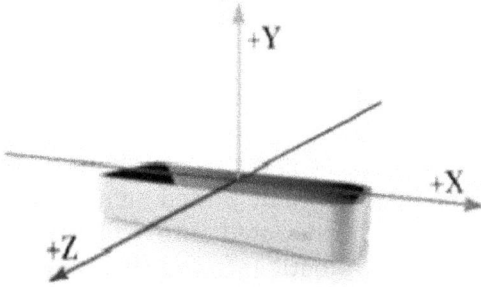

Figure 2. Leap Motion coordinate system.

Methodology and Design

In the proposed system, an input gesture is acquired using a Leap Motion sensor. It tracks hands and fingers movements in 3-D digital format. It gives a mapping of the gesture in terms of feature points. The extracted fingertip positions of each gesture are stored in the database. The distance between them is used as feature vector. While testing, again gestures are captured, and the positions of the finger points are extracted. Distances are calculated by Euclidean distance method for current gesture and all gestures stored in the database. For recognition, the feature vectors are compared using the four similarity measures viz. Euclidean distance measure, Cosine similarity, Jaccard Similarity and dice similarity. The one with maximum similarity is returned as the detected gesture [7-12].

Functionality of the framework begins with user writing a character in air, the Leap Motion Controller gets track the movements of the finger. After pre-processing takes place frames are generated and feature extraction is done. Using computer vision techniques, the text written in air is recognized on the basis of strokes. The recognized text is then displayed on the screen. For the evaluation of the proposed framework some experiments were created. Those experiments were carried out by some students. Each student was asked to write English alphabets (Upper case or Lower case), numeric digits or a complete word in the air. One by one each student starts writing in the air in front of LMC. The proposed framework starts recognizing the alphabets and numeric digits. The recognition accuracy and the number of attempts were recoded and

17

analyzed. The results obtained from these experiments were presented and discussed in the following section [13-15].

Leap Motion Controller

The Leap Motion controller is a small peripheral device shown in Figure 3, which is designed to be placed on a physical desktop, facing upward. It can also be mounted onto a virtual reality headset. It consists of two monochromatic Infrared (IR) cameras and three infrared LEDs, the device observes a roughly hemispherical area, to a distance of about 1 meter. The LEDs generate patternless IR light and the cameras generate almost 200 frames per second of reflected data. This is then sent through a USB cable to the host computer, where it is analyzed by the Leap Motion software using "complex math" in some way synthesizing 3D position data by comparing the 2D frames generated by the two cameras.

Data Description

Leap Motion Controller (LMC) is being used as input device capable to record finger gestures perform within the interaction region. The input data is for example, user writing in air with fingers or performing gestures with hands. All these hand movements are being tracked by Leap Motion Controller and are being sent to our application. To calibrate initially the LMC with the computer through hands, we've fixed few gestures. For air writing the finger movements is being tracked and recordsthe strokes. Finger movements for different alphabets in English, numeric, words and some especial gestures for erase and space are recorded as input data. There are 20 people are taken to record the input data to get the variations in writing styles.

Strokes

Stroke is sudden or sharp change in the movement, or it is a mark made by drawing a figure in one direction. Mathematically or geometrically, it is the maximum curvature in a line curve. Here are some examples of the strokes recorded against different gestures generated for writing letters or words in English. The alphabet 'A' has three strokes. The system stores the strokes template, which is used for recognition process. Following figure show the examples of the hand written alphabets, numeric and word ('Casper') in English.

Figure 3. Handwritten alphabets – Different stroke style of alphabets A and Z.

After the strokes the corresponding data about all the information of strokes is then taken from inspect element. Following template shows an example of the data stores against the stroke. Due to large size of the template, digits are omitted, and few are inserted instead. The recognition of the alphabet or the text uses the template which is displayed on the screen accordingly. There cognition process of alphabets and numeric digits is taking place on basis of strokes; an expert system is being used to recognize the letters [16-19].

Training

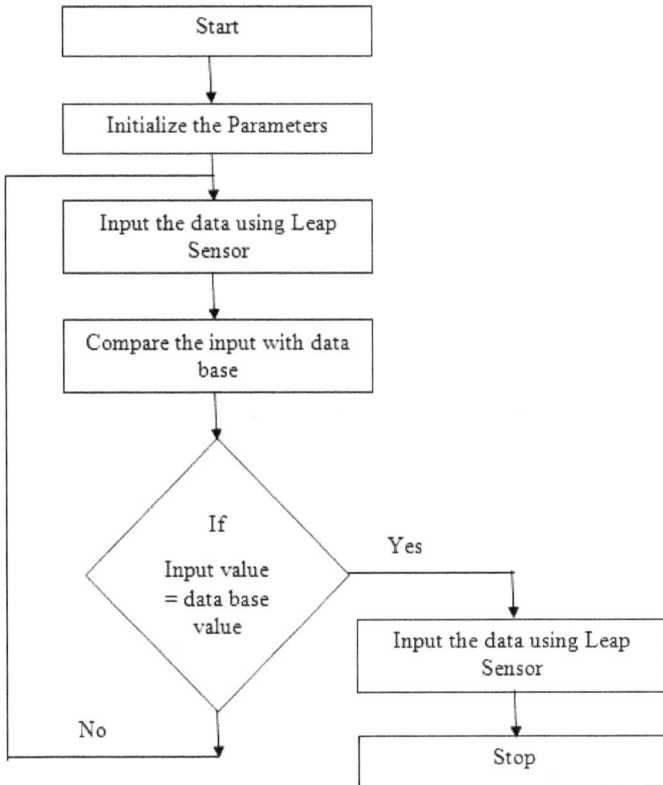

Figure 4. System performance - Flow chart

We had a significant data corresponding to each of alphabet in English in addition to numeric therefore we motivated to use Neural Network and Fuzzy Logic to train the system for the proposed framework. Neural Network is a field of Artificial Intelligence which finds data structures and algorithms for learning and data classification which is inspired by a normal human brain. Neural Network techniques can be used to learn through examples and create a structure of rules to classify the different kinds of inputs for example the recognition of images. Fuzzy logic is a computing-based approach which is based on 'degrees of truth' rather than the usual 'true and false' of '1 or 0' Boolean logic on which most of the modern computers are based on. In Fuzzy Logic the true value can be

20

anywhere in between 0 and 1 and 0 and 1 are extreme cases of truth. In the proposed framework neural network has been used to train the model. The output of the trained model was based on the Fuzzy logic instead of binary classification [20-22]. System performance model is shown in Figure 4.

Results and Discussion

Table 1. Observations of average number of attempts for recognition of small alphabets

Input Data	Average Accuracy	Number of Attempts
a	91	10
b	95	8
c	97	5
d	94	7
e	94	7
f	95	5
g	92	10
h	93	10
i	97	4
j	95	5
k	96	5
l	99	2
m	94	6
n	95	6
o	90	3
p	92	7
q	93	8
r	92	7
s	95	6
t	92	6
u	93	7
v	92	6
w	90	10
x	90	10

y	92	10
z	93	8

The average numbers of attempts for English alphabets, numeric digits, special characters and some words were also analyzed. Table 1 shows the observations of average number of attempts for recognition of small letter English alphabets and the accuracy results obtained after testing is mentioned in Figure 5. The framework was getting little confused in the recognition of "O", "0" and "Q" for these characters we got accuracy result of 90%, 93% and 88% because of the resemblance in their shape. We have recorded the results of each character (capital alphabets, small alphabets, special characters, numeric values and different words) in the tables below.

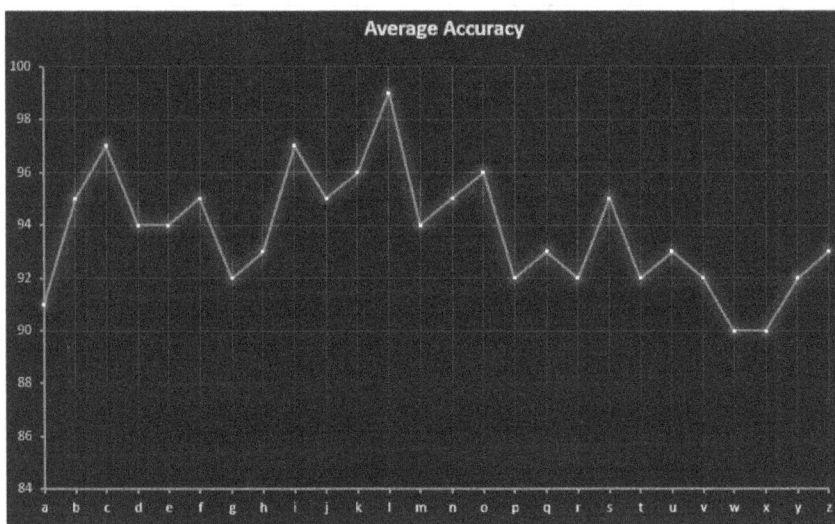

Figure 5. System accuracy for small alphabets.

Table 2 shows the observations of average number of attempts for recognition of capital letter English alphabets and the accuracy results obtained after testing is mentioned in Figure 6.

Table 2. Observations of average number of attempts for recognition of capital alphabets

Input Data	Average Accuracy	Number of Attempts
A	94	8
B	91	10
C	95	9
D	92	9
E	91	10
F	93	9
G	92	10
H	94	9
I	97	5
J	96	3
K	94	5
L	93	7
M	89	11
N	91	10
O	90	8
P	92	10
Q	87	15
R	86	15
S	94	6
T	90	8
U	93	12
V	92	12
W	89	15
X	91	10
Y	90	10
Z	95	9

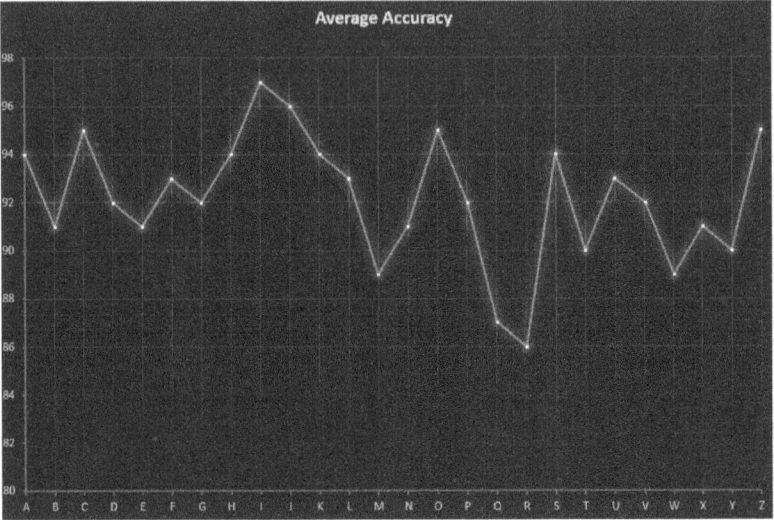

Figure 6. System accuracy for capital alphabets.

Table 3 shows the observations of average number of attempts for recognition of numeric digits and the accuracy results obtained after testing is mentioned in Figure 7.

Table 3. Observations of average number of attempts for recognition of numeric digits

Input Data	Average Accuracy	Number of Attempts
0	91	6
1	99	5
2	95	6
3	94	7
4	92	7
5	91	7
6	94	6
7	95	6
8	91	7
9	92	6

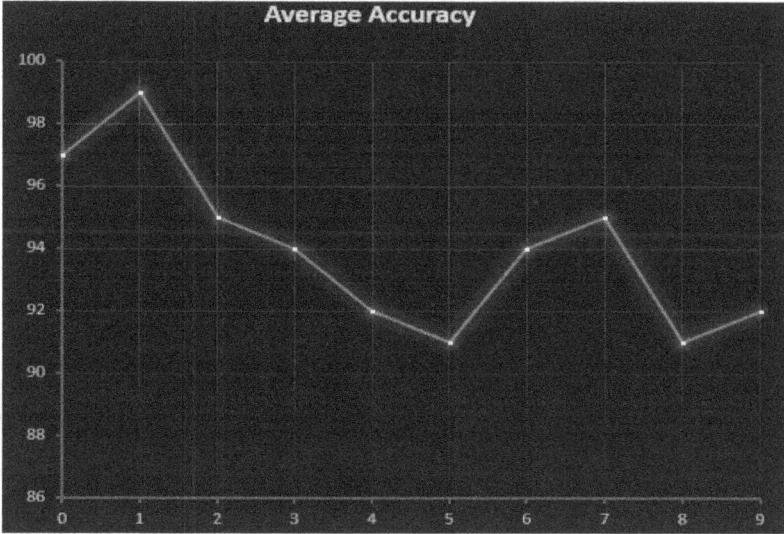

Figure 7. System accuracy for numeric digits.

Table 4 shows the observations of average number of attempts for recognition of special characters and the accuracy results obtained after testing is mentioned in Figure 8.

Table 4. Observations of average number of attempts for recognition of special characters

Input Data	Average Accuracy	Number of Attempts
μ	84	10
α	86	10
β	80	15
α	88	15
θ	88	15
λ	89	15
%	78	20
&	84	15
@	86	15
#	90	10
?	86	15

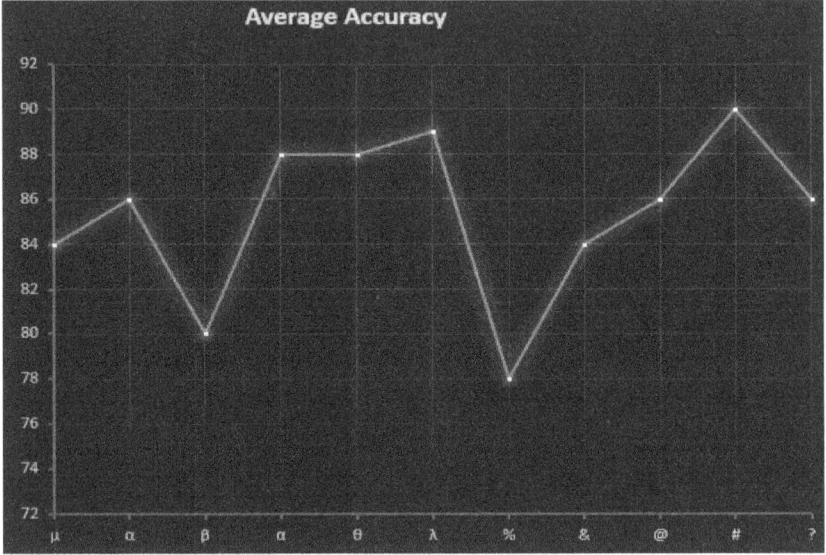

Figure 8. System accuracy for special characters.

Table 5 shows the observations of average number of attempts for recognition of small letter English alphabets and the accuracy results obtained after testing is mentioned in Figure 9.

Table 5. Observations of average number of attempts for recognition of words

Input Data	Average Accuracy	Number of Attempts
hi	82	15
hello	76	15
how	78	15
who	79	15
by	85	15
name	74	15
AND	76	15

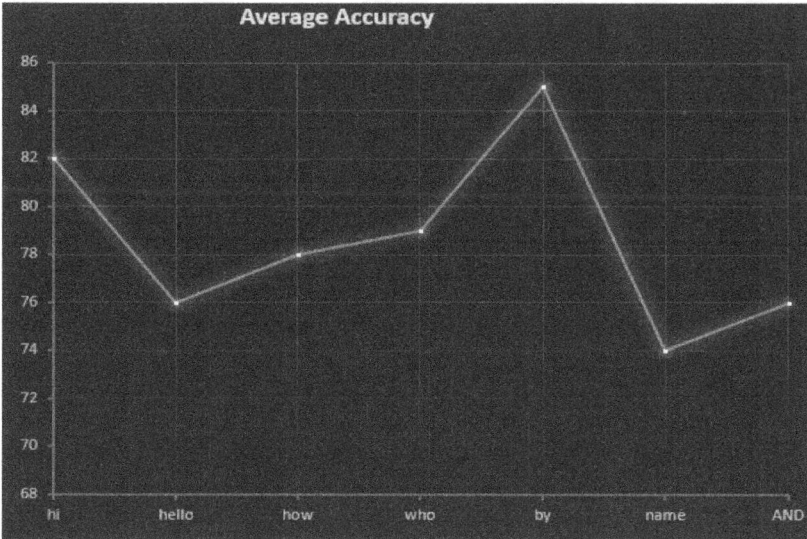

Figure 9. System accuracy for words.

Leap Motion Controller is being used as input device capable to record finger gestures perform within the interaction region. The input data is for example, user writing in air with fingers or performing gestures with hands. All these hand movements are being tracked by Leap Motion Controller and are being sent to the application. Then the text written in air is recognized on the basis of strokes and recognized text is then displayed on the screen. Experimental set up and different strokes are shown in Figure 10.

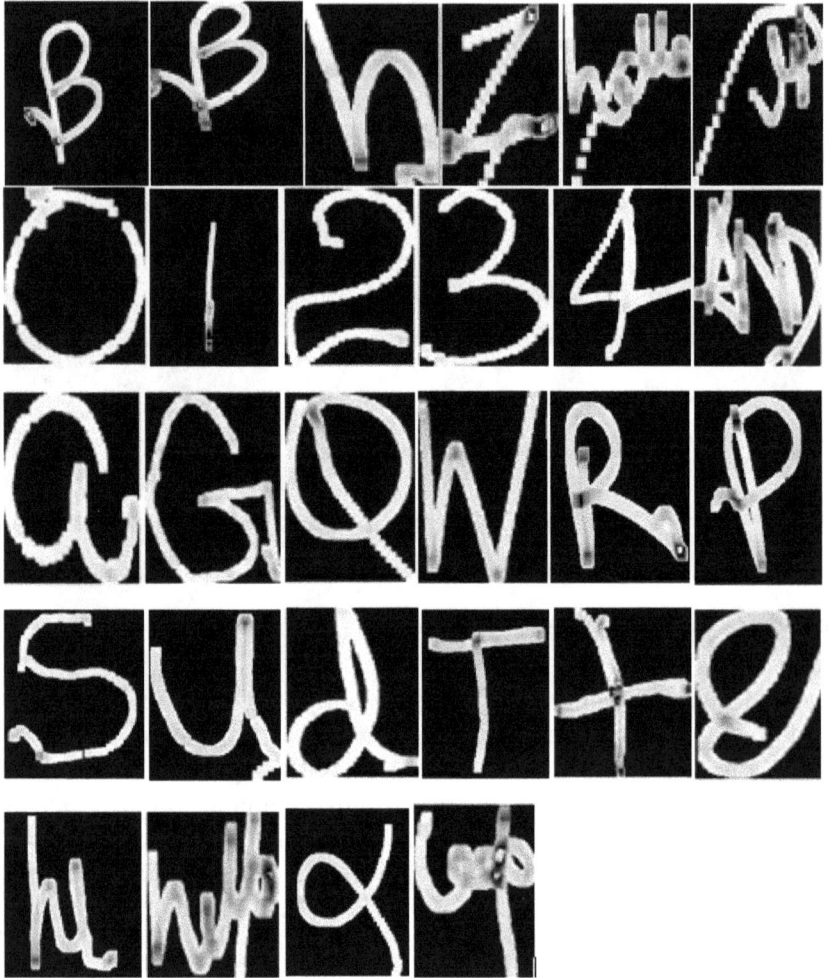

Figure 10. Experimental setup.

The proposed system architecture is shown in Figure 11, and the system performance is shown in Figure 12.

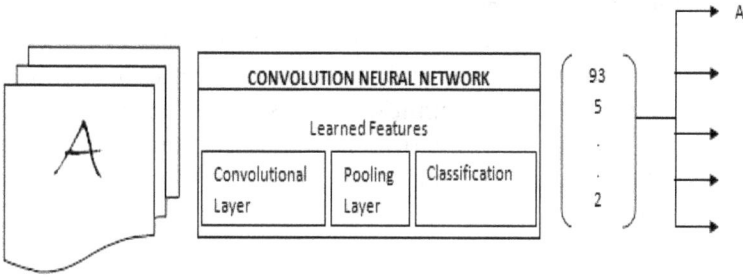

Figure 11. CNN.

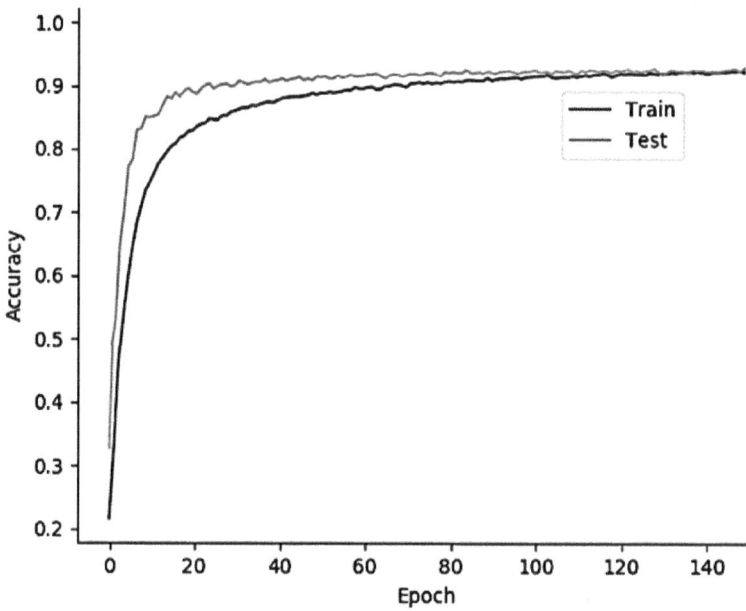

Figure 12. Accuracy of proposed neural network.

Conclusion

This presented framework which enables its user to literally write in the air through his/her figure and hand movement. The framework records strokes of the gestures and apply computer vision approach to recognize written characters. These characters could be single English alphabets, numeric or a complete connected word. Experimental results demonstrated high accuracy in recognition. This framework will also help to do basic physiotherapy for hand or finger which is proposed future direction of this framework. The evaluation of each alphabet separately and for most of the characters (upper- and lower-case alphabets and numeric digits) the system shows the average recognition rate is greater than 90%. In future this system can be transformed to new system which can enable dyslexic write alphabets naturally in air and to make their learning experience fun.

References

1. Luo, Y.; Tsang, C.C.; Zhang, G.; Dong, Z.; Shi, G.; Kwok, S.Y.; Li, W.J.; Leong, P.H.W.; Wong, M.Y. An Attitude Compensation Technique for a MEMS Motion Sensor Based Digital Writing Instrument. In Proceedings of the 2006 1st IEEE International Conference on Nano/Micro Engineered and Molecular Systems, Zhuhai, China, 18–21 January 2006; pp. 909–914.
2. Nguyen, H.; Bartha, M.C. Shape writing on tablets: Better performance or better experience? In Proceedings of the Human Factors and Ergonomics Society, Boston, MA, USA, 22–26 October 2012; pp. 1591–1593.
3. Amma, C.; Georgi, M.; Schultz, T. Airwriting: A wearable handwriting recognition system. Pers. Ubiquitous Comput. 2014, 18, 191–203.
4. Kratz, S.; Rohs, M. Protractor3D. In Proceedings of the 15th International Conference on Intelligent User Interfaces—IUI '11, Palo Alto, CA, USA, 13–16 February 2011; ACM Press: New York, NY, USA, 2011; p. 371.
5. Frolova, D.; Stern, H.; Berman, S. Most Probable Longest Common Subsequence for Recognition of Gesture Character Input. IEEE Trans. Cybern. 2013, 43, 871–880.
6. De, O.; Deb, P.; Mukherjee, S.; Nandy, S.; Chakraborty, T.; Saha, S. Computer vision based framework for digit recognition by hand gesture analysis. In Proceedings of the 2016 IEEE 7th Annual Information Technology, Electronics and Mobile Communication Conference, Vancouver, BC, Canada, 13–15 October 2016; pp. 1–5.
7. Poularakis, S.; Katsavounidis, I. Low-Complexity Hand Gesture Recognition System for Continuous Streams of Digits and Letters. IEEE Trans. Cybern. 2016, 46, 2094–2108.
8. Qu, C.; Zhang, D.; Tian, J. Online Kinect Handwritten Digit Recognition Based on Dynamic Time Warping and Support Vector Machine. J. Inf. Comput. Sci. 2015, 12, 413–422.
9. Hwang, H.J., Kim, S., Choi, S. & Im, C.H. (2013) EEG-Based Brain-Computer Interfaces: A Thorough Literature Survey, International Journal of Human-Computer Interaction, 29:12, 814-826
10. Pavaloiu, I.B., Petrescu, I. & Dragomirescu, C. (2014) Interdisciplinary Project-Based Laboratory Works, The 6th International Conference Edu World 2014 "Education Facing

31

Contemporary World Issues", 7th - 9th November 2014, Procedia - Social and Behavioral Sciences, Volume 180, pp. 1145–1151

11. Bachmann, D., Weichert, F. & Rinkenauer, G. (2015) Evaluation of the Leap Motion Controller as a New Contact-Free Pointing Device, Sensors 2015, 15, 214-233.

12. Weichert, F., Bachmann, D., Rudak, B. & Fisseler , D. (2013) Analysis of the Accuracy and Robustness of the Leap Motion Controller, Sensors 2013, 13, 6380-6393.

13. Silberman, N.; Fergus, R. Indoor Scene Segmentation Using a Structured Light Sensor. In Proceeding of IEEE International Conference on Computer Vision Workshops (ICCV Workshops), New York, NY, USA, 6–13 November 2011; pp. 601–608.

14. Chen, F.; Brown, G.; Song, M. Overview of three-dimensional shape measurement using optical methods. Opt. Eng. 2000, 39, 10–22.

15. Aslan, et al. "Mid-air authentication gestures: an exploration of authentication based on palm and finger motions." Proceedings of the 16th International Conference on Multimodal Interaction. ACM, 2014.

16. G. Bailador, et al. "Analysis of pattern recognition techniques for in-air signature biometrics." Pattern Recognition 44.10 (2011): 2468-2478.

17. K. M. Vamsikrishna, et al. "Computer-Vision-Assisted Palm Rehabilitation With Supervised Learning." IEEE Transactions on Biomedical Engineering 63.5 (2016): 991-1001.

18. C. Agarwal et al. "Segmentation and recognition of text written in 3d using leap motion interface." Pattern Recognition (ACPR), 3rd IAPR Asian Conference on. IEEE, 2015.

19. M. Piekarczyk et al. "On using palm and finger movements as a gesture-based biometrics." Intelligent Networking and Collaborative Systems (INCOS), International Conference on. IEEE, 2015.

20. C. Barber, et al. "The quickhull algorithm for convex hulls." ACM Transactions on Mathematical Software (TOMS) 22.4 (1996): 469-483.

21. P. Kumar, et al. "3D text segmentation and recognition using leap motion." Multimedia Tools and Applications (2016): 1-20.

22. P. Kumar, et al. "3D text segmentation and recognition using leap motion." Multimedia Tools and Applications (2016): 1-20.

Chapter 3

Design and Analysis of Microstrip Antenna for UWB Applications

Piyush Mishra, Prateek Goel, Priyanshu Yadav, Rajan Kumar Singh and Sarabjeet Kaur
Electronics and Communication Engineering Department, NIET Greater Noida Uttar Pradesh, India

Abstract

This paper exhibits the plan and investigation of microstrip patch antenna for ultra- wideband applications which are portrayed through recreations in HFSS. Here antenna works in a frequency band 4.3GHz to 11.6GHz including C band and X band. This antenna is reasonable for wireless computer networks.

Introduction

This paper gives a brief layout of the UWB antenna design. The business recreation device is utilized to show how antenna size and shape impact the UWB antenna boundaries. In view of simulation, a technique to further develop UWB boundaries is proposed. Microstrip patch antennas are fairly planar antenna which have been audited in latest forty years. Microstrip patch turns into the top picks among antenna designer with numerous applications in military just as in business areas. The possibility of microstrip patch antenna came from printed circuit innovation for circuit components and correspondence lines as well as for transmitting parts of electronic framework. The key improvement of the microstrip patch antenna includes region of metallization maintained over a ground plane by a slim dielectric substrate and dealt with against the ground at a fitting region. The patch shapes can be rectangular, three-sided, circle, ring etc. The metallic patch is typically comprised of meager copper foil. The substrate material gives mechanical help to the exuding patch parts. It like savvy keeps the fitting degree between the patch and its ground plane. The substrate thickness for the essential calculation is in the scope of 0.01 to 0.05 wavelength. By and large utilized material is Teflon- based with an overall permittivity somewhere in the range of

2 and 3. This material is likewise called PTFE (Polytetrafluoroeth-ylene). It has a design extremely indistinguishable from fiberglass material utilized for advanced circuit sheets, however has a much diminished misfortune digression. In this antenna having fractional ground has been planned whose patch is a mix of rectangular and semi- roundabout patches which reverberates at various frequencies. This is a ultra- wideband antenna with a repeat extent of 4.3-11.6 GHz with double resounding frequencies which covers both C and X groups frequencies appropriate for wireless communications. The substrate utilized in this antenna is FR4. The proposed antenna is limited with irrelevant production cost. To design and meet the ideal presentation particulars of microstrip patch antenna, the antenna originator should secure a blend of specific insight and abilities. In the first place, he ought to have a comprehension of the essentials of activity of the fundamental microstrip patch antenna engineering. This can be cultivated from an actual model known as the cavity model, which depends on various suppositions appropriate to thin substrates. By and by, it is strange that the fulfillment points of interest can be met by the fundamental microstrip patch antenna design. Thick substrates and added features, as parasitical patches, shorting pins, or spaces in the patch, should be added.

Results and Discussion

The limit of the proposed printed antenna was affirmed and upgraded by really open Technology like HFSS simulator. The patch antenna was made on the printed circuit board for useful insight, as depict in Figure. Notwithstanding little component of proposed antenna, it can change over a huge bandwidth transmission that covering the full reach from 4.45 to 7.1 GHz that is allotted for UWB application. The addition just as productivity was impacted by the lossy FR4 dielectrics materials that was utilized as a substrate material.

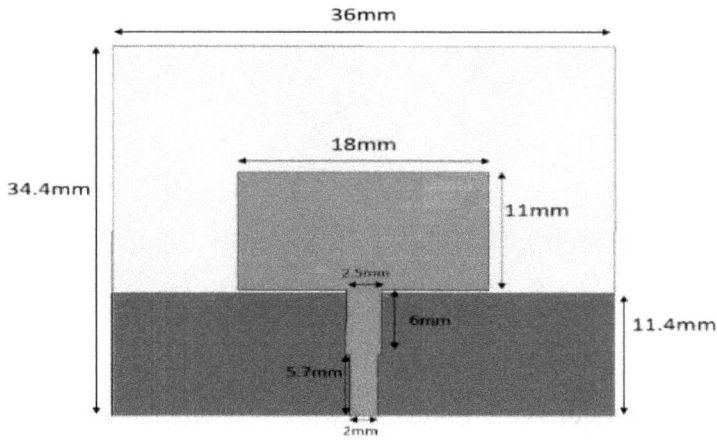

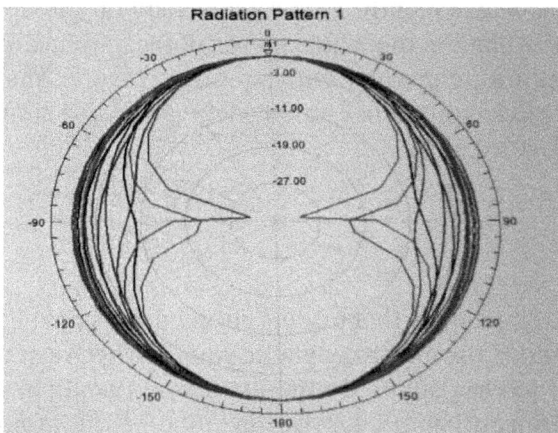

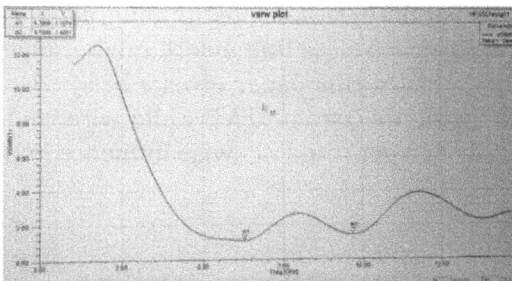

Using even more costly microwave substrate diverge from standard low worth FR4, the increment and capability of the proposed patch antenna can be updated.

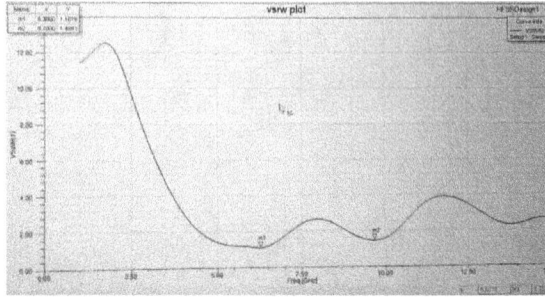

Current dispersion is more steady in a lower band than with an upper band. In upper band, the formation of an electric field close to openings is sensible. Subsequently, excitation is solid on the upper band just as the lower band in the whole pieces of the antenna. The deliberate outcomes are comparable with simulated results. The outcome might show a tad inconsistency among estimation and simulation in light of manufacture errors.

Conclusion

Simulated outcomes show that placing corner scores on the radiating patch which builds the electromagnetic coupling between the transmitting patch and the awful ground plane and results into the upgraded data transmission. Likewise inclining of the ground plane makes smooth change from one resonant frequency to other; which has additionally further increased impedance transmission bandwidth. Gotten radiation design almost omnidirectional. Henceforth impedance to existing wireless framework can be stayed away from. The antenna presents in this paper gives a brief considered Ultra-Wide Band. Changing the size of UWB radio wire could further develop radiation design yet could likewise influence impedance coordinating.

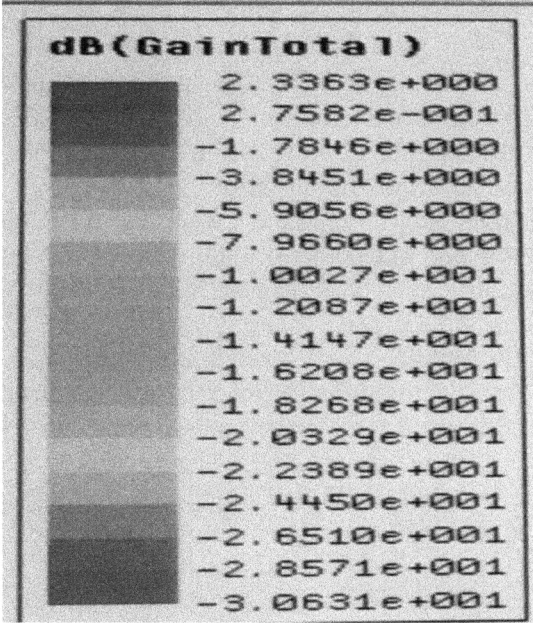

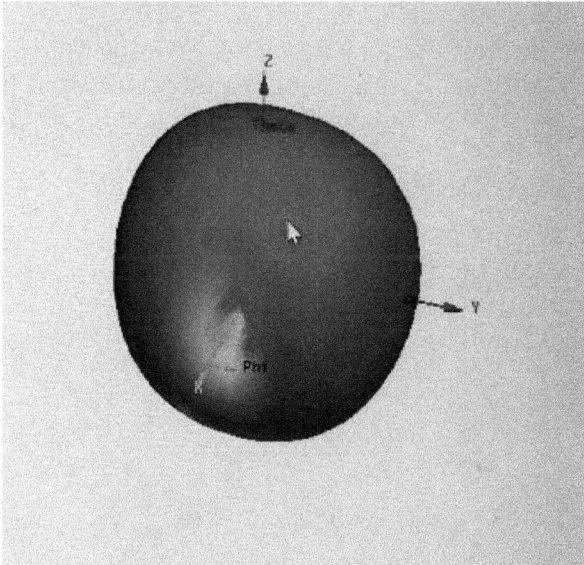

References

[1] Zengrui Li, Zhun Yang, Qingxin Guo, Junhong Wang, Wenbo Jiang, "Printed triple-t monopole antenna for 2.4, 5.2 and 5.8GHz WLAN operations", International Conference on Microwave Technology and Computational Electromagnetics, 2009

[2] A. A. Deshmukh, P. Mohadikar, K. Lele, P. Verma, D. Singh and K. P. Ray, "Ultra-wideband star shaped planar monopole antenna," IEEE International Conference on Advanced Networks and Telecommunications Systems (ANTS), pp. 1- 6, 2016

[3] Amit A. Deshmukh, Payal Mohadikar, Priyanka Verma, Priyal Zaveri, K. P. Ray, "Ultra-wideband compact ring Sector printed monopole antenna", International Conference on Computing Communication Control and automation (ICCUBEA), pp. 1-6, 2016

[4] Jang Hwan Bae, Jun Gi Jeong, Young Joong Yoon, Yongwook Kim, "A compact monopole antenna for bluetooth and UWB applications", IEEE, ISAP 2017

[5] A. Hachi, H. Lebbar, M. Himdi, "A novel small and compact flexible monopole antenna for UWB applications", International Conference on Optimization and Applications (ICOA), 2018.

[6] Abdullah Haskou, Anthony Pesin, Jean-Yves Le Naour, Ali Louzir, "Compact, Dual-Band, Hybrid Monopole-ASA, Antenna", IEEE International Symposium on Antennas and Propagation, July 2019.

Chapter 4

Refreshable Electronic Braille Display

R.A. Dhejas Vaishnavi, K. Ganga Devi and K. Immanuel M.E.
Electronics and Communication Engineering, St. Joseph's Institute of Technology, Chennai, India

Abstract

Braille System is a method that is widely used by visually impaired people to read and write. Braille Code generally consists of cells of raised dots arranged in a grid to inscribe characters on paper. Blind people can sense the presence and absence of dots using their fingertips, giving them the code for symbol. It is difficult for visually impaired people to receive visual information. The main objective of this project is to establish a means of communication for them. Here we are implementing Enhanced Braille System that helps blind people to read text. We scan image from camera, process the image by image processing techniques and that will be converted into text. The detected text is given to the Raspberry Pi, which recognizes every character and convert it into Braille Code. Here solenoids are used to facilitate the vertical movement of the dots of the Braille Display which is controlled by Raspberry Pi. With this device, we are displaying that Braille Code on the Braille Keypad.

Introduction

Braille is a writing cum reading method used by blind or visually impaired people. In braille script, each character is represented by a unique combination of 6 dots, which form one cell. It has always been a challenge to make braille books for blind people. On an average a page of normal text takes around 4 pages in braille which often makes braille books quite bulkier which involve a higher production cost. So, to work around this problem we have come up with a solution to make a braille display which can take text as well as image input and convert that into braille. There are currently many refreshable braille displays available in the market but all of them face a major problem, i.e., high cost, so we propose a design in order to tackle thischallenge.

Proposed System

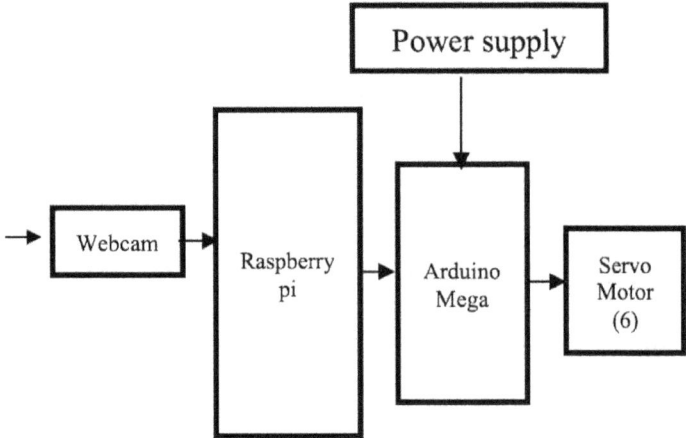

Figure 1. Block diagram of the proposed system.

The proposed sensor had different layers of elastomers as epidermis-dermis, piezo resistive cantilevers were placed behind the ridge which were responsible for unique deformation of the sensor. The proposed sensor was quite efficient but needed a lot of precision for scanning the exact angle. Increasing Braille Literacy: Voice-Assisted Electronic Braille Books (Mahmoud Al-Qudsi et al,2013) (eBraille eBook) for the Visually Impaired this paper aims at making eBooks to braille through a braille display plus an additional voice assistance this is similar to out text input, where direct text can be given as input to the file. So, here we propose a design that that solves the above challenges. The text to brailledisplay that we have designed uses a SBC (Raspberry Pi) and an Arduino board which can convert any text or image to braille. A raspberry pi is a SBC, with great computational power. Raspberry Pi is a credit card sized single board computer designed by Raspberry Pi Foundation, which can beplugged into a monitor or TV. It was primarily designed for teaching coding to kids of elementary school, but overtime pi has built its reputation as a Multitasking Board which is widely used by various hobbyists and makers for their projects.

Algorithm

```
┌─────────────────────────┐
│   CAPTURE THE IMAGE     │
└───────────┬─────────────┘
            │
            ▼
┌─────────────────────────┐
│   CONVERT INTO TEXT     │
└───────────┬─────────────┘
            │
            ▼
┌─────────────────────────┐
│ COMMUNICATION BETWEEN   │
│   TWO CONTROLLERS       │
└───────────┬─────────────┘
            │
            ▼
┌─────────────────────────┐
│ SERVO MOTOR RUN BASED   │
│       ON IMAGE          │
└─────────────────────────┘
```

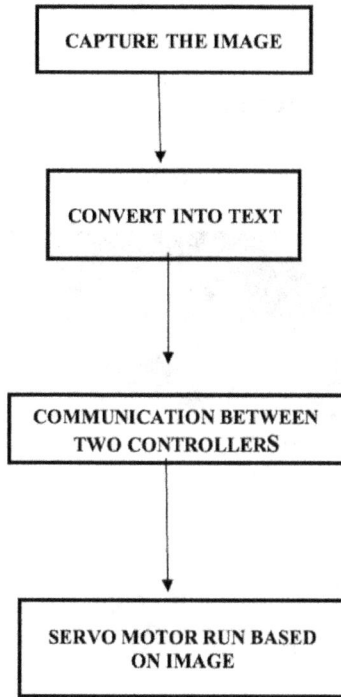

A. Capture Image and Convert to Text

The Webcam is used to capture the image and send the image to raspberry pi. Tensor flow package is used to identify the object and convert to text mode. The converted text will be obtained one by one to create a product name.

B. Communication Between Two Controllers

The raspberry pi controller has a converted text file. Then they transmit the data to Arduino mega through the serial communication. The Arduino receiver part receive the data from the master device.

C. Servo Motor Run Based on Image

The braille board have six dots. So, we are using 6 servo motor to

43

make the board electronically. The Arduino mega give the instruction for servo motor to run the text-based pattern. The braille board have a default pattern for every alphabetic character. Based on the received character the servo motor pattern will change.

Experimentation and Results

Capturing the Image

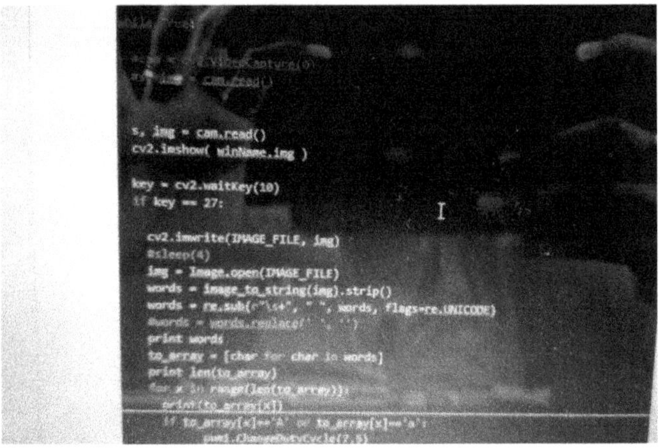

Figure 2. Capturing the image.

Image to Text Conversion

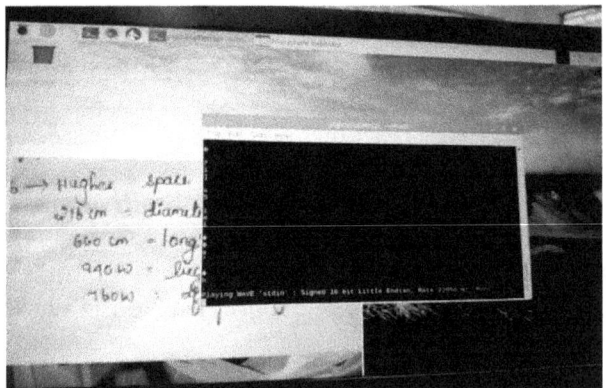

Figure 3. Image to text conversion.

We make use of Pytesseract OCR library to read and analyze the text embedded in the printed textbook, the function Image_to_string() helps us in returning the text after running OCR. The extracted image is printed and written in a new .txt file. Each character issegregated individually by use of a for loop and square bracket function. Each character is given unique matrix code that corresponds to that character'sactual Braille dot arrangement. By using a For loop and an If-elif- else loop we get the output for each character individually. The code is printed sequentially; since Python doesn't have an inbuilt Matrixsupport we get the output in form of lists thatcan be considered as Matrices.

Conclusions

This image to braille convertor can be added with a camera module and hence the images can be captured in a regular time interval and then the images can be represented as text on the refreshable display, allowing the blind people to read normal books and journals comfortably. Moreover, they need not wait for the braille version of the book or journal to come out. Our model of the refreshable display has a speed of representing the braille characters at a huge rate of 12 character in 2.5 seconds and on an average 290 characters in 60 seconds. Our complete system works on just a power supply of 15.4 Watts. The raspberry pi here is poweredby a 5 V, 2 A charger in which pi is performing the processing and 2 servo motors are powered through pi. The Arduino is connected with a 9 V, 0.6 A charger, which helps Arduino produce PWM signals and 4 servo motors are powered from the Arduino board.

Figure 4. Setup of the proposed system.

The image is captured and saved in the default library, the Picamera module is called through its function PiCamera(). The capture() function is used to capture the image and save it in the specifiedlocation.

References

[1]   A. S. Al-Fahoum, H. B. Al-Hmoud, and A. A. Al- Fraihat, "A smart infrared microcontroller-based blindguidance system," Act. Passive Electron. Compon., vol. 2013, pp. 1–7, Jun. 2013. [14] A. A. Nada, M. A. Fakhr, and
A. F. Seddik, "Assistive infrared sensor based smart stickfor blind people," in Proc. Sci. Inf. Conf. (SAI), Jul. 2015,pp. 1149–1154.
[2]   R. Dhod, G. Singh, G. Singh, and M. Kaur, "Low costGPS and GSM based navigational aid for visually impairedpeople," Wireless Pers. Commun., vol. 92, no. 4, pp. 1575–1589, Feb. 2017.
[3]   S. Khan, K. Ahmad, M. Murad, and I. Khan, "Waypointnavigation system implementation via a mobile robot usingglobal positioning system (GPS) and
global system for mobile communications (GSM) modems," Int. J. Comput. Eng. Res., vol. 3, no. 7, pp. 49–54, 2013.
[4]   R. Tapu, B. Mocanu, and T. Zaharia, "Wearable assistive devices for visually impaired: A state of the art survey," Pattern Recognit. Lett., vol. 137, pp. 37–52, Sep.2020.
[5]   S. Keele, "Guidelines for performing systematic literature reviews in software engineering," EBSE, Goyang-si, South Korea, Tech. Rep. Ver. 2.3, 2007.
[6]   B. Kitchenham, R. Pretorius, D. Budgen, O. P. Brereton,

M. Turner, M. Niazi, and S. Linkman, "Systematic literature reviews in software engineering-a tertiary study," Inf. Softw. Technol., vol. 52, no. 8, pp. 792–805, 2010.

[7] B. Kitchenham and S. Charters, "Guidelines for performing systematic literature reviews in software engineering version 2.3," Engineering, vol. 45, no. 4ve, p.1051,2007. [8] B. Kitchenham, Procedures for PerformingSystematic Reviews, vol. 33.Keele, U.K.: Keele Univ., 2004, pp. 1–26.

[9] S. Khan, A. Hafeez, H. Ali, S. Nazir, and A. Hussain, "Pioneer dataset and recognition of handwritten pashto characters using convolution neural networks," Meas. Control, Nov. 2020, Art. no. 002029402096482, doi:10.1177/0020294020964826.

[10] S. Nazir, S. Khan, H. U. Khan, S. Ali, I. Garcia-Magarino, R. B. Atan, and M. Nawaz, "A comprehensive analysis of healthcare big data management, analytics and scientific programming," IEEE Access, vol. 8, pp. 95714– 95733, 2020.

[11] A. Hussain, S. Nazir, S. Khan, and A. Ullah, "Analysis of PMIPv6 extensions for identifying and assessing the efforts made for solving the issues in the PMIPv6 domain: Asystematic review," Comput. Netw., vol. 179, Oct. 2020, Art.no. 107366.

[12] S. Nazir, S. Shahzad, and N. Mukhtar, "Software birthmark design and estimation: A systematic literaturereview," Arabian J. Sci. Eng., vol. 44, no. 4, pp. 3905–3927, Apr. 2019.

Chapter 5

Pet Daycare Bot with Autofeedingand Remote Surveillance Using Raspberry pi

D. Elamathy, K. Gayathri and A. Dinesh Kumar M. E.
Electronics and Communication Engineering, St. Joseph's Institute of Technology, Chennai, India

Abstract

Household pets need special treatment and care. They need to be attended to as at when due with food. Due to busy life style of most owners, this task may not be as simple as expected. Lack of adequate attention to pets' needs might have great consequential effects, such as starvation, ill health, among others. In view of the foregoing, this work proposes an Internet of Things based automated feeder system that uses Raspberry pi to drive its remote control, scheduling and intelligence. Its design and subsequent implementation is expected to, at least, take care of the nutritional aspects of pets by providing as either scheduled or intelligently the food and drinks of pets as at when due in the absence of the owner. Thus, this work aims to automate the monitoring and feeding process that is usually done manually by pet owners. To achieve the foregoing, the proposed system uses a food dispenser that is connected to a microcomputer which is programmed to control the feeder as scheduled, remotely or intelligently. Thus, allowing the user to have full control over the time a pet is fed and the amount of food consumed by the pet. The feeder can be controlled through a secure web application hosted on a local server and through advance scheduling. Our study not only presents the key improvement of the pet monitor system involved in the ideas of the Internet of Things, but also meets the demands of pet owners, who are out for works without any trouble. The objective is to allow pet owners to automate simple things, like monitoring, and feeding controls.

Introduction

Internet of things is the coming together of internet and physical devices in a network of unlimited possibilities using microcontrollers, arduino and raspberry pi. IOT allows for physical devices to wirelessly communicate over networks which has led to a growing number of applications for iot devices. People love their pets and vice versa, but there are times you need to leave your pets at home for long durations alone and this is a problematic issue. Well we here propose to design a dog daycare robot that can monitor as well as feed the dogs or cats in a timely manner. Remote Monitoring of your pets over IOT. Speaking to pets by Voice Commands.2x Containers: For Dog/Cat Food and Water. Timely Feeding of Cat/Dog Food and Water through Feeding Tray. Main Modules of Robot Well Protected from Naughty Pets using Steel Mesh.

Proposed System

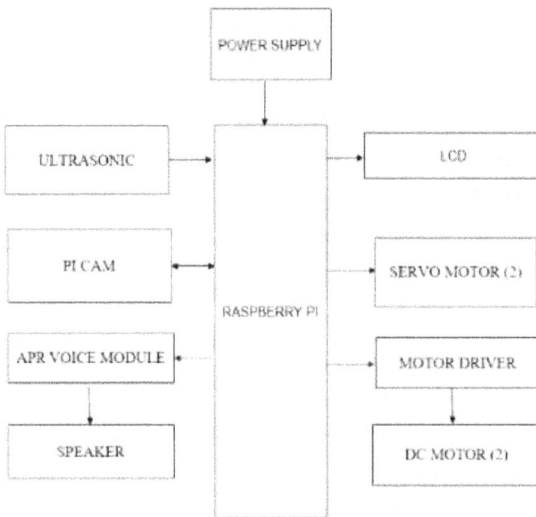

Figure 1. Block diagram of the proposed system.

In the existing system variety of dog feeders are available but it does not have remote surveillance facility, even it is available the dog might get scared by the robot.

In the proposed system the bot is integrated with a camera that allows for live streaming over IOT platform that allows you to monitor and control the pets online. We can control the pet even when we are not at home the chase move towards the dog calls out

50

the pet like an owner when the dog is nearer it automatically opens its tray for feeding food. The whole setup is easily controlled through the website by the owner. Here the voice of the owner is stored in the voice module, so the pet does not get scared, and it also enjoy the company of robot like an owner. The dog also eats the food properly, even if it damages the things in home, the owner can control them easily by giving commands to their pet lively. The robot finds applications in feeding any kinds of pet from any-where, remotely explore the house for any potential theft and de-ceive robbers to believe that someone is at home. It also includes personalize daily meal portions, stay connected with real time alerts, know your pet is alright when you are away, stores up to 7lbs, keeps food fresh and takes cares of your pet's diet. The robot also enables a user to prevent a pet from eating a specific food while still allowing access to that food to other pets. All these fea-tures make the robot user friendly for users having more than one pet.

Algorithm

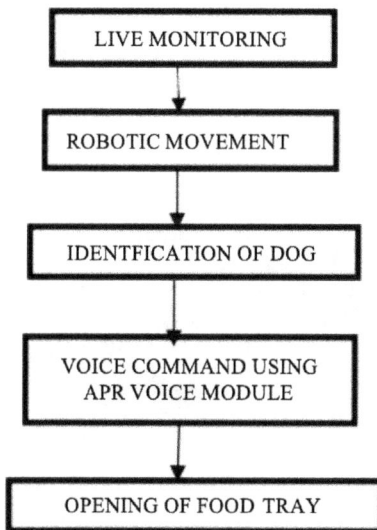

Figure 2. Flowchart of pet daycare robot.

Live Monitoring

Using picam, the pet can be completely monitored through internet by the owner when they are not in home, and the data is streamed. It is very useful for the owner to monitor their dog and their actions.

Robotic Movement

The movement of motor driver is controlled by owner through the webpage, it moves in every direction and follows the dog like an owner.

Identification of Dog

Using ultrasonic sensor, the robot detects the distance of the dog and get stopped when it gets alert at certain range.

APR Voice Module

Here the owner speaks with their pet the speaker is interconnected by the APR voice module, the voice of the owner is also stored in it.

Food Tray

With the help of servo motor the food tray open and close when it detects the dog on the feeding time.

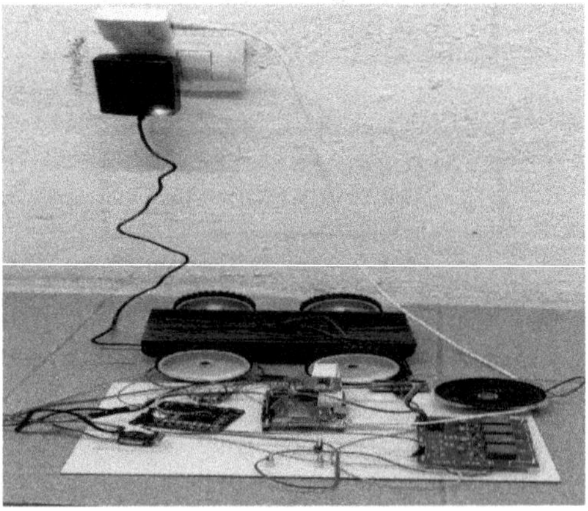

Figure 3. Experimental setup.

Experimentation and Results

The first part is the live monitoring of dog through the picam attached to the robot, which continuously streams the data to the owners phone.

Figure 4. Live streaming.

The owners can control the movement of chase setup and move them towards the dog.

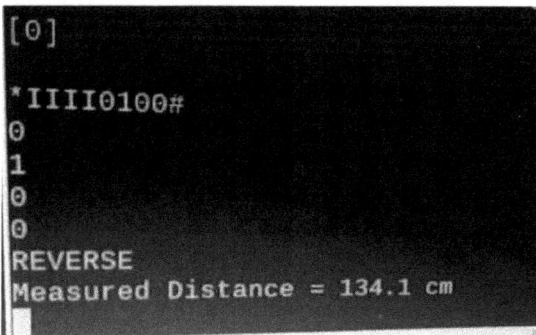

Figure 5. Controlling the movement.

The movement may be forward, reverse, left or right based on the owners wish. Secondly, it detects the dog or other things using ultra sonic sensor, getting alert, so it protects the setup from damaging by hitting on the wall or other things.

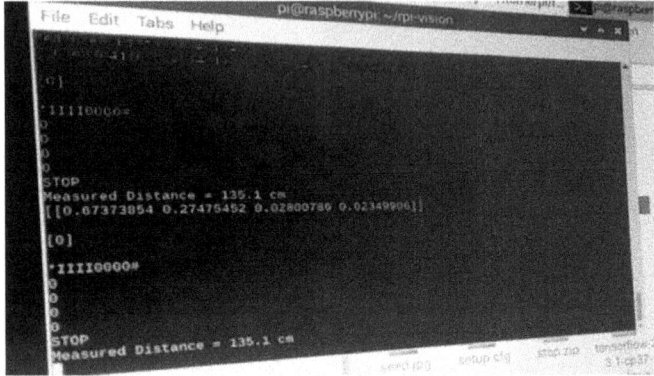

Figure 6. Alert message.

Finally, if it detects the dog the owner can give commands to the dog through the APR voice module, which makes the pets not to feel lonely.

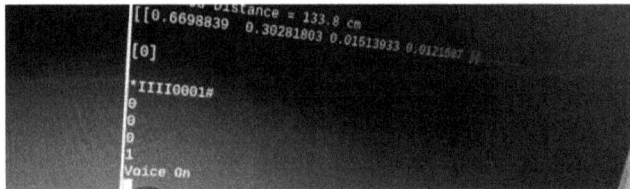

Figure 7. Voice module.

The owner can call out their pet through the voice module at the time of food. So, the dog gets fed properly even when the owners are out of station. The pet daycare robot has lot of advantages, it will be very useful for the pet lovers.

Conclusions

This project brought together several components and idea to achieve a common goal   that is design an automated pet feeder using Raspberrypi. the key components of the project include a servo motor sensor which will be programmed to serve the food as soon as the pet comes. It relieves the owner from having to feed his pet multiple times a day. The proposed project senses the presence of the pet using the WIFI which is the consumers can watch their pets through the smartphone and serves accordingly. The owner does not have to worry about making plans or feeding his pet because of this pet daycare robot. This automatic pet feeder serves as a helping hand as it works efficiently in the absence of the own. Through this project, it helps to develop creativity in crating projects and modify existing projects to be more energy efficient with new fabrication methods. This pet daycare robot makes it easy for consumers to feed their pets and will not leave their pets hungry again. The effectiveness of the project used while feeding pets requires picam to make it easier for users to see their pets inside the mobile device. This indicates that the project has the potential to be extended to external agencies to expand its use. It is recommended that the promotion to be carried out for commercialization purpose.

Figure 8. Setup of the proposed system.

ACKNOWLEDGEMENT

At the outset, we would like to express our sincere gratitude to our beloved Chairman B. Babu Manoharan for his constant guidance and support. We would like to express our thanks to B. Jessie Priya and B. Shashi Sekar for their kind encouragement and blessings. We express our sincere gratitude and whole-hearted thanks to P. Ravichandran for his encouragement to make this project a successful one. We wish to express our sincere thanks and gratitude to C. Gnana Kousalya for leading us towards the completion of this project. We also wish to express our sincere thanks to A. Dinesh Kumar for her guidance and assistance in solving the various intricacies involved in the project. Finally, we thank our parents and friends who helped us in the successful completion of this project.

References

[1] K. Li, Z. Zhang, W. Liu, Automatic Test Data Generation Based On Ant Colony Optimization, International Conference on Natural Computation. W. R. Penrose, J. R. Stetter, Detection of Contraband Food Products with a Hybrid Chemical Sensor System, IEEE, (2021)1073-1076.

[2] D. J. L. Cuesta, D. R. Suarez, and L. H. C. Gordo, Sistema Dosificador de Solidos para Alimentacion de Perros con Modulo de Comunicacion Remoto Solid Dosing System for Feeding Dogs with Remote Communication Module, IEEE, (2021) 1-5.

[3] M. Pardo, L. G. Kwong, G. Sberveglieri, J. Schneider, W. R. Penrose, J. R. Stetter, Detection of Contraband Food Products with a Hybrid Chemical Sensor System, IEEE, (2021) 1073-1076.

[4] Martini B G, Helfer G A, Barbosa J, et al. "Indoor Plant:A Model for Intelligent Services in Indoor Agriculture Based on Context Histories". Sensors, 2021, 21(5):1631.

[5] Liu C, Li J, Balamurugan S, et al. "Cognitive computing for intelligent robots in assisting preschool children". Intelligent Service Robotics, 2020(4).

[6] Ren S, He K, Girshick R, et al. "Faster R-CNN: Towards Real-Time Object Detection with Region Proposal Networks". IEEE Transactions on Pattern Analysis and Machine Intelligence, 2015, 39(6).

[7] Hu Jie, Shen Li, Albanie Samuel, et al. Squeeze-and-Excitation Networks. 2020,42(8):2011-2023.

[8] Hu W P, Lin C B, Yang C Y, et al. "A Framework of the Intelligent Plant Factory System". Procedia Computer Science, 2018, 131:579-584.

[9] Hybrid Chemical Sensor System, IEEE, (2021)1073-1076. [14] S. Subaashri, M. Sowndarya, D. K. S. Sowmiyalaxmi, Automatic Pet Monitoring and Feeding System Using IoT, International Journal

of Chem TechResearch, 10(14) (2017) 253-258.

[10] F. Han. J, Yang, Y. Liu, and H. Zhao, Research on Preprocessing Algorithm for PET - CT Image Registration of MR – based Attenuation Map in PET/MR, IEEE, (2017)1-3.

[11] Vineeth S, Sneha Lakshmi V C, Prashant Ganjihal, Rani B, "Review on Development of Automatic Pet Food Dispenser using Digital Image Processing., SSRG International Journal of Electronics and Communication Engineering 6(11)(2019) 6-8.

[12] Y. Shi, and B. Yu, Output Feedback Stabilization of Networked Control System with Random Delays Modeled, IEEE, 54 (7) (2009) 1668-1678.

[13] L. C. Lin and T. B. Gau, Feedback Linearization and Fuzzy Control for ConicalMagnetic Bearings, IEEE, 5 (4)(1997) 417-426.

Chapter 6

IOT based Smart Sericulture System using Arduino

T. Lakshmi, S. Madhumitha and S. Sathiya Priya
Electronics and CommunicationEngineering, St. Joseph's Institute of Technology, Chennai, India

Abstract

Sericulture is the heart of raising silkworms for silk production. India is the second largest producerof silk in the world. Sericulture is the root of social, economic, cultural and political progress of India. Temperature and humidity play an important role in the development of healthy silkworms in every stage, especially during the development of larva. In our project, we dispense a MEGA controller to plot the real time because to observe or keep track of the silkworms. Image processing is helpful in recognizing the infection or ill health and non-identical stages of the silkworms. Our specimen keeps up the collection of the real time statistics by using MEGA controller. The total organization or structure is statistics and execute with help of MEGA controller. In identical stages of the silkworm the MEGA is check or control the atmospheric environment or surrounding inside the room of the silkworms rearing. Minimize the manual intervention of the farmer by automating the process of irrigation of mulberry plantation and also testing the temperature and controlling the silkworm rearing unit by using MEGA board. Image processing technique mainly used to find out the colour change in the silkworm's body. It indicates the non-identical stagessuch as black worms and swallow worms indicates the diseases worms.

Introduction

Sericulture refers to the cultivation of Silkworms for the production of Silk. Parameters like Temperature, Humidity, Soil moisture and Light intensity are the important factorsin the progression of silkworms and suitable encouraging must be done according to the requisities in every stage. Environmental variations assume as the important part in the growth and development of silkworm. Sericulture an important occupations, but the techniques used by the agriculturists are outdated. Hereafter, there is a need of developing modernization in sericulture cultivation. This endeavor gives a thought of providing automation in sericulture production.

The model goals at making use of developing technology that is IoT and Smart Sericulture using automation. Observing environmental parameters of the silkworm rearing house is the most important aspect to improve the vintage of the silk. The specialty of this model comprises enhancement of a system which can observe temperature, humidity, light power through sensors using NodeMCU and in case of any variation in the parameters send a notification on the users' mobile application using internet connection. This system permits for data assessment and scheduling to be programmed through theArduino IDE software.

ASIA is the main in production of silk which produces over95% of the total global output with bulk production of it in China and India, followed by countries like Japan, Brazil, USA, Italy and Korea. Sericulture is an art and science raising silkworm for silk production. Silkworms are Stenophagous insects that are fed mulberry leaves and/or silkworm chow. In the adult phase of the life cycle, the silkworm moths do not eat or drink. There are four different stages namely egg, larva,pupa, and moth. Sericulture activities are broadly classified into two; the agro-based sector and the industrial sector. The agro-based part involves two distinct phases of activities that is mulberry cultivation and silkworms rearing. Silkworm rearing is differentiated into two stages: young age from 1st and 2nd instar. This intermediate stage will be the 3rd instar and the rearing of 4th and 5th instars comes under late age rearing.

The sensors network utilized in our smart sericulture system comprises of smart sensor nodes interfaced with temperature and humidity sensors to collect real time accurate readings inside the system. The auto-controlled actuators namely, exhaust fan, heater and sprinkler maintain the temperature and humidity of the system within the threshold levels. Image processing technology is utilized to capture the pictures of silkworms and to analyze the status of sericulture process.

Day by day silkworm production is decreasing because the farmers facing so many problems by following the traditional method of sericulture; hence, we collected information related to healthy growing of silkworms and it require environmental conditions. They shared the problems which they are facing due to the traditional sericulture method. Literature survey helped us to select better ideas and technology to implement our project.

Literature Review

In 2017, Amandeep, Arshia Bhattacharjee, Spandan Ghosh, Sayan Saha Souvik "Smart Farming Using IoT" Even today, different developing countries are also using traditional methods and backwards techniques in agriculture sector. Little or very less technological advancement is found here that has increased the production efficiency significantly. To increase the productivity, a novel design approach is presented in this paper. Smart farming with the help of Internet of Things (IOT) has been designed. A remote-controlled vehicle operates on both automatic and manual modes, for various agriculture operations like spraying, cutting, weeding etc. The controller keeps monitoring the temperature, humidity, soil condition and accordingly supplies water to the field [1].

In 2016, Divya Darshini B, Adarsh B U, Shivayogappa H J "Automated Smart Sericulture System based on 6LoWPAN and Image Processing Technique"Sericulture is the heart of raising silkworms for silk production. Temperature and humidity play an important role in the development of healthy silkworms in every stage, especially during the development of larva. Disinfection is one of the critical parameters to be considered for healthy and successful silkworm rearing. In this paper, we present an 6LoWPAN (IPv6 over Low power Wireless Personal Area Network) enabled IoT based approach to design a real time sericulture monitoring and disinfection actuating system with an inclusion of image processing technology to identify the stages of silkworm life cycle. Our prototype supports real time data collection using emerging 6LoWPAN, CoAP and RPL protocol. The complete system is designed and implemented using Contiki OS to control the atmospheric condition inside the sericulture system as per the requirements in each stage of sericulture life cycle. This complete prototype was built using the TelosB motes running 6LoWPAN stack interfaced to temperature and humidity sensors with an disinfection actuation system and a serial camera to auto capture the pictures and to analyze it using an

61

image processing method to check the status on sericulture process [2].

In 2015, M A Dixit, Amruta Kulkarni, Neha Raste, Gargi Bhandari "Intelligent Control System for Sericulture (the manufacture of silk) is an important rural occupation. India is the world's second largest silk producer, accounting for around 15% of global production after China, which accounts for a staggering 80%. An examination of Indian sericulture processes reveals an obvious need for automation, particularly during the pre- cocoon stages. During this phase, the silkworms go through critical bodily changes that impact the quality and amount of the silk produced. Maintaining optimal abiotic parameters such as temperature, humidity, and other conditions results in a significant increase in the quantity and quality of silk output. An intelligent sericulture plant automation system that employs zone-based cascade control can be one the solution. The actuator sub-system uses the actuators in that zone of the unit to implement the remedial actions. Continuous real-time input makes it easier to make precise and timely corrections. The technique is designed to boost silk quantity and quality, which is governed by reeling factor, holding capacity, and silk roughness. Furthermore, the zone-based implementation lowers production and maintenance costs, making it suited for use in rural areas.[3].

In 2017 Kamilaris, A. Kartakoullis, and F. X. Prenafeta- Boldu, "A review on the practice of Big data analysis in Agricultuture" The complex agricultural ecosystems must be better understood in order to address the growing problems of agricultural production. This is possible because to modern digital technologies that continuously monitor the physical environment and generate massive amounts of data at an unparalleled rate. Farmers and businesses would be able to extract value from this (big) data through analysis, increasing their productivity. Despite the fact that big data analysis is advancing different industries, it has yet to be broadly adopted in agriculture. The purpose of this study is to review current agricultural studies and research projects that use the new practice of big data [4].

In 2017 F. K. Shaikh, S. Zeadally, and E. Exposito, "Enabling technologies for green Internet of Things" The carbon footprint has increased as a result of recent technological advancements. Over the last several years, energy efficiency in the Internet of Things (IoT) has gotten a lot of attention from researchers and designers, paving the way for a new field termed green IoT. To allow a green IoT environment, efficient energy utilisation is required in several parts of IoT (such as key enablers, communications, services, and applications). We investigate and debate how various enabling technologies (such as the Internet, smart items, sensors, and so on) may be used to establish a green IoT. We also look at a variety of IoT applications [5].

In 2015 Perara, C. H. Liu, and S. Jayawardena, "The Emerging Internet of Things marketplace from an industrial perspective: A survey" The Internet of Things (IoT) is a dynamic worldwide information network made up of Internet-connected devices such as RFID tags, sensors, actuators, and other instruments and smart appliances that are becoming an important part of the future Internet. We've seen a lot of IoT solutions developed by start-ups, small and medium businesses, large corporations, academic research institutes (such as universities), and private and public research organizations make their way into the market over the previous decade. In this paper, we look at over a hundred IoT smart solutions on the market and analyze them to see what technologies, features, and applications they use. We divide and discuss these solutions into five groups based on the application domain [6].

Methodology

Temperature sensor is used to detect the heat of the silkworms' room, and if temperature increases the exhaust fan is ON to cool the temperature. Humidity sensor is used to sense the content of moisture present in air. LDR is mainly used to monitor the intensity and brightness of light. The LCD is interface with arduino to display the results of temperature, humidity, moisture sensor and air quality sensor.

Parameter	Requirement
Temperature	20°C to 30°C
Humidity	70% to 90%
Light intensity	15 to 30 LUX (dim light)

Air quality	Not more than 1% (Avoidpolluted air)

Design and Implementation

In the proposed method, Arduino UNO microcontroller is used to interface with the sensors. The DHT11 sensor used to monitor the room temperature in silkworms area. Whenever the temperature is high the exhaust fan will run and reduce the temperature.

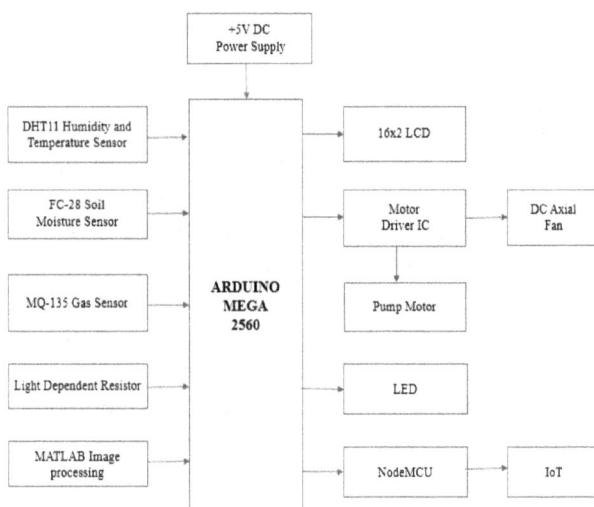

Figure 1. Block diagram of smart sericulture system.

The soil moisture sensor is used to detect the moisture level in the soil. Based on the moisture level present in thesoil the pump motor will run and pump the water to the mulberry plants. The MQ-135 air quality sensor is used to monitor the air quality present inside the room. The LDR (Light Dependent Resistor) is used to detect the time day or night. Based on LDR condition the bulb will turn ON/OFF. The MATLAB processing is used to detect the silkworm condition. If the silkworm is in the unhealthy condition the medi-cine pump motor will turn on and spraythe medicine. The LCD is used to display all the information. If any unhealthy silkworms are identified in the captured image, then the sprayer will turn ON and spray some medicine to the silkworms.

Efficient wireless sensor network with IOT technology to monitor and control the temperature, humidity and light intensity present in silkworm rearing house. No need high manpower. Gainful employment, economic development and improvement in the quality of life to the people in rural areas.

Result Analysis

Image processing system includes treating images as two-dimensional signals while applying already set signal processing methods to them. Identify the colour change in the body of the worms , which indicates the different stages and the light yellowish indicates that they have reached to the test proves that implemented prototype is successfully capable to monitor the parameters in real time and to control the condition inside the deployed environment and has several advantages in term of remote monitoring, automated actuation to suitable condition inside the system, image processing to know the real time status in complete sericulture process, low cost of the system, flexibility, user friendliness and energy efficiency.

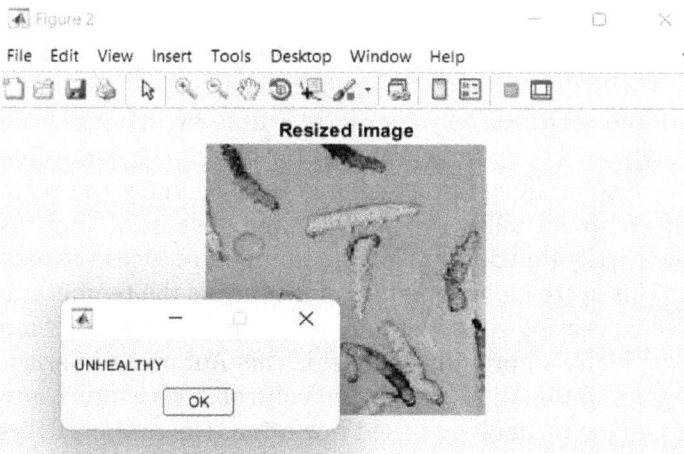

Figure 2. Result 1- Matched image is unhealthy worms.

Figure 3. Result 2- Matched image is healthy worms.

Image processing using Deep learning (CNN algorithm) is used to identify the healthy and unhealthy silkworms in a real time system.

Conclusion and Future Enhancement

Our project presents a design of smart monitoring and actuating sericulture system using Arduino mega and Image processing technologies. The Arduino mega will enable the end user to monitor and to actuate the sericulture system in real time by making use of an iot. Prototype will operate in real time for monitoring and actuation inside the system. Image processing is used to know the real time status of complete silkworm growth and diseases. This system avoids over irrigation and topsoil erosion and reduces the wastage of water. This system also monitors and controls the temperature of the silkworm rearing room continuously, within the fixed temperature range, hence improving the yield. This Automated Sericulture System" gives automation and guided control in sericulture advances by employing technology-based invention. The proposed system facilities and conduct the environmental conditions to be reserved inside the silkworm rearing house. Required edge values for parameters like temperature, relative humidity and light intensity can be stable based on the environmental circumstances. On the basis of requirement fan, light, and heater is turned on and off based on required environmental condition. The planned system is financially affordable and power effective organization. Implemented test of

66

this prototype system validates that the proposed system can work gradually to observe the environmental conditions inside the silk-worm raising house. The proposed system reduces the manpower and reduces the chance of errors. The model is easy to implement and use. Using various deep learning algorithm to identified the affected silkworm and classified what kind of disease will be affected and automatically prescribed the medicine for the particular people. It recommends financially affordable and power effective organization.

References

[1]   GayathriNagasubramanian, Rakesh Kumar Sakthivel, Rizwan Patan, Muthuramalingam Sankayya, Mahmoud Daneshmand, Amir H.Gandomi, "Ensemble Classification and IoI-Based Pttern Recognition for Crop Disease Monitoring System", IEEE IoT Journal, Vol.8, no. 16, pp.12847-12854, 2021.

[2]   Othmane Friha, Mohamed Amine Ferrag, Lei Shu, Leandros Maglaras, Xiaochan Wang, "Internet of Things for the Future of Smart Agriculture:A Comprehensive Survey of Emerging Technologies", IEEE/CAA Journal of Automatica Sincia, vol.8, no.4, pp.718-752,2021.

[3]   S.D. Manikandan, Sadhish Prabhu, ParnasreeChakraborty, T. Manthra, M. Kumaravel, "IoT-Based Smary Irrigation and Monitoring System in Smart Agriculture", Futuristic Communication and Network Technologies, vol.792, pp.45, 2022.

[4]   Harshit Bhatt, Brij Bhushan Sharma, Aaditya Sharma,Ruchika Sharma, Meenakshi Sharma, "An IoT- BasedSoil Properties Monitoring System for Crop Growth and Production", Privacy and Security Challenges in Location Aware Computing, pp.200, 2021.

[5]   Roopashree, Kanmani, Babitha, Pavanalaxmi, "Smart Farming with IoT:A Case Study", Internet of Things and Analytics for Agriculture, Volume 3, vol.99, pp.273, 2022.

[6]   Ashish Gupta, Hari Prabhat Gupta, Preti Kumar, Rahul Mishra, Surbhi Saraswat, Tanima Dutta, "A Real=time Precision Agriculture Monitoring System using Mobile Sinkin WSNs", 2018 IEEE International Conference on Advanced Networks and Tele-communicationSystem (ANTS), pp.1-5, 2018.

[7]   Uferah Shafi, Rafia Mumtaz, Syed Ali Hassan, Syed Ali Raza Zaidi, Awais Akhtar, Muhammad Moeez Malik, "Crop Health

Monitoring Using IoT-Enabled Precision Agriculture", IoT Architectures, Models and Platforms for Smart City Application, pp.134, 2020.

[8]   Perara, C. H. Liu, and S. Jayawardena, "The EmergingInternet of Things marketplace from an industrial perpective: A survey," IEEE Trans. On Emerging Topics in Comp., vol. 3, no. 4, pp, 585-598, Dec 2015.

[9]   F. K. Shaikh, S. Zeadally, and E. Exposito, "Enabling technologies for green Internet of Things," IEEE System J., vol. 11, no. 2, pp. 983-994, 2017.

[10] Kamilaris, A. Kartakoullis, and F. X. Prenafeta-Boldu, "A review on the practice of Big data analysis in Agricultuture," Comp. and Elec. In Agri., vol. 143, pp. 23- 37, 2017.

[11]   L Ricardo de Queiroz, "Processing JPEG-Compressed Images and Documents", IEEE Transaction on image Processing, vol. 7, no. 12, December 1998.

[12]   V K Rahamathulla, "Management of Climatic Factors for Successful Silkworm (Bombyxmori.) Crop and Higher Silk Production: A Review" in, Hindawi Publishing Corporation Psyche, vol. 2012, pp. 12.

[13]   Guobao Xu, Weiming Shen and Xianbin Wang, "Application of Wireless Sensor Networks in Marine Environment Monitoring: A Survey" 1424-8220.

[14]   M S Sunita, Jyoti Malik and Suman Mor, "Comprehensive Study of Application of Wireless Sensor Network", International Journal of Advanced Reasearch in omputer Science and Software Engineering, vol. 2, no. 11, November 2012.

[15]   B R Patil, K K Singh, S E Pawar, L Maarse and J Otte,"Sericulture: An Alternative Source of Income to Enhance the Livelihoods of Small-scale Farmers and Tribal Communities", Pro-Poor Livestock Policy Initiative A Living from Livestock Research Report, pp. 09-03, July 2009.

Chapter 7

Portable Ventilator And HealthcareMonitoring System Using Arduino

P. Princy Magdaline, R. Kutti Raja and B. Lakshman
Electronics and CommunicationEngineering, St. Joseph's Institute of Technology, Chennai, India

Abstract

We usually consider of air pollution as being outdoors, but the indoor air has also been polluted. Indoor air quality problems have caused problems for human body. The world is facing the health issues and the inherent need for assisted-living environments for citizens.There is also a commitment by National Healthcare Organizations to increase support for personalized, integrated care to prevent and manage chronic conditions. Respiratory diseases are pathological conditions which makes gaseous exchange harder due to the malfunction of tissue and organs responsible for breathing. Two very common examples of respiratory disease include: Chronic Obstructive Pulmonary Disease (COPD) and Obstructive Sleep Apnea (OSA). Many applications related to In-Home Health Monitoring have been introduced over the last few decades, thanks to the advances in mobile and Internet of Things technologies and services. The proposed system is based on the Internet of Things technology and Embedded system. The system also includes electronic devices and sensors. The main objective of this system is smart ventilator with health care monitoring system using micro controllers and sensors based on IoT. With the help of high torque motor, we can change the pressure value up to three levels. The sensors collect the values and update it in the IoT module. The data is stored in a web page which can be accessed from anywhere. The data also helps in patient's progress analysis.

Introduction

Amid the global crisis caused by the corona virus pandemic, hospitals and healthcare facilities are reporting shortages of vital equipment. As makers it's our responsibility to combat the shortage by constructing makeshift-open-source substitute devices. IoT technologies have matured since its conception a decade ago, with increasingly successful implementations at smart city and smart home projects around the world.

One important device for which demand has ramped up is ventilators for patients who need assistance with their breathing due to the respiratory effects of COVID-19. Basically, a ventilator is a machine that provides breathable air into and out of the lungs, to deliver breaths to a patient who is physically unable to breathe or breathing insufficiently. A DIY ventilator may not be efficient as that of a medical grade ventilator, but it can act as a good substitute if it has control. Human lungs use the reverse pressure generated by contraction motion of the diaphragm to suck in air for breathing. A contradictory motion is used by a ventilator to inflate the lungs by pumping type motion. A ventilator mechanism must be able to deliver in the range of 10 – 30 breaths per minute, with the ability to adjust rising increments in sets of 2. Along with this the ventilator must have the ability to adjust the air volume pushed into lungs in each breath. We have also operated the ventilator manually and remotely through the help of IoT. We can also monitor the patient's heath continuously and store the data in webpage. So that the doctor can access the data from anywhere.

Literature Review

In 2021, Nada Y. Philip, Joel J. P. C. Rodrigues, Simon James Fong, Honggang Wang, Jia Chen, worked on "Internet of Things for In-Home Health Monitoring Systems: Current Advances, Challenges and Future Directions". With the growing society, traditional healthcare systems reach their capacity in providing sufficient and high-quality services. The world is facing the aging population and the inherent need for assisted-living environments for senior citizens. There is also a commitment by national healthcare organizations to increase support for personalized, integrated care to prevent and manage chronic conditions. It allows healthcare providers to reach patients outside of the four walls of the hospital, perform proper monitoring of patient health conditions, continue to deliver quality care and identify at-risk populations. It also helps patients stay connected with their health providers, enable them to remain compliant with treatment plans and improve their health conditions.[1].

In 2021,Mashoedah , Umi Rochayati , Indra Hidayatulloh, Arya Sony, Ferda Ernawan Fardiansyah, Apri Nuryanto Worked on "IoT Enabled Ventilator Monitoring System for Covid-19 Patients ".The use of the IoT protocol on medical equipment is expected to provide protection for medical personnel in dealing with Covid-19 patients, especially when medical personnel are monitoring and setting up an equipment. This study aims to produce a monitoring and control system for a breathing apparatus (Ventilator) based on the Internet of Thing (IoT), test the ventilator control function, test the data transmission function with the IoT protocol. The method used is Define, Design, Develop, and Disseminate (4D). Data collection is done through Testing and Observation Limited field test. This research produces a control and monitoring system for mechanical ventilators. The role of the ventilator is vital and is needed in the handling of covid19 patients. However, with a limited amount, not all patients can be treated using this equipment. Asreported by several print or electronic media, transmission of covid19 virus is through droplets (small droplets) that come out of the patient's body and transmission can also be through the air. Covid19 transmission mechanism like this requires us to maintain a physical distance (physical distancing) ofat least 1-2 meters when talking with others.[2].

In 2020, Sivapriya Natarasan,Dr.Pavithra Sekar worked on "Design and Implementation of Heartbeat rate and SpO2 Detector by using IoT for patients ".In this modern days, development of smart healthcare system is an emerging area of research on the Internet of Thing (IoT). The majority of the individuals who are living in provincial territories are not able to get medical services because of the absence of specialists, emergency and private clinics. Likewise, notwithstanding a little medical issue, the individuals get wavered to get counsel from a specialist due to travel, cost, and time. The heart is the most vital organ in the human body, and the heartbeat rate is an essential aspect of human metabolism. Pulse oximetry is the non-invasive measurement of the oxygen saturation (SpO2) in the heart. However, most of the available heart rate measurement tools are rather expensive and only available in hospitals. The vital monitoring system is a non invasive technique and heartbeat rate, spo2 and temperature detector with an android mobile app. This

71

monitoring system measures heartbeat rate, SpO2 and temperature which uses SpO2 probe for measuring oxygen saturation and heart-beat rate which also includes temperature probe to measure body temperature. The probe has to place on the person's fingers to measure the concentration of the HbO2 and Hb. For measuring, the pulse oximetry was used which uses lights to measure the oxygen saturation.[3].

In 2020, Subha R, Haritha M , Nithishna B , Monisha S G worked on "Coma Patient Health Monitoring System Using IoT". This proposed system equipment has been structured utilizing Arduino controller board. In subsumption different sensors canbe associated through the controller board interface. At The fundamental goal of this task is to build up an online application to correspondence with a web server in a verified way for coma patients. Constant and progressing association with persistence is the best approach to constrain or propel others to take comparing activities. Here the information can be seen in both PC and mobile. The sensors worth will be transferred in an incorporated cloud server. The proposed system uses a heterogeneous IoT plan to verify correspondence between a sensor hub and an Internet hub and this plan is indistinct against different conditions. Numerous analysts have contemplated different research issues on incorporating IoT wellbeing observing and sensor advances. the Doctor to keep a keen eye on the patient's health. So a system is used to monitor the overall health status of a patient, which needs constant care, the data at receiver which can be used to analyze the patients overall health condition. At present, endeavors are being made to incorporate these two advancements on the equivalent IoT stage in various fields. Dissimilar to customary investigations that give IoT stages at the design level just, this examination proposed a usage model of a sensor information store on the SQL Server. [4].

In 2017, R.Harini , B. Rama Murthy worked on "Development of a Wireless Blood Pressure Monitoring System by Using Smartphone". Telemedicine system is used for patient monitoring as well as diagnosis of diseases of remote area patients. Telemonitoring is a medical practice that involves remotely monitoring patients who are not at the same location as the healthcare provider. In emergency case of patient continuous monitoring of vital signs is necessary one of the most important vital sign is blood pressure. When your heart beats, it contracts and pushes blood through the arteries to the rest of your body. This force creates pressure on the arteries. Blood pressure is recorded as two numbers, the systolic pressure over the diastolic pressure (as the heart relaxes between beats). The proposed system objective is to develop a blood pressure monitoring device controlled by Arduino microcontroller. The software implementation is in a form of android app application. The proposed system attempts to design and implementation of patient monitoring in real time with wireless transmission via Wi-Fi. B.P meter sensor,microcontroller & Android technology to transmit data wireless in Smartphone, as great use in the field of medicine and helps the Doctor to keep a keen eye on the patient's health. So, a system is used to monitor the overall health status of a patient, which needs constant care, the data at receiver which can be used to analyze the patients overall health condition. The systolic, diastolic & pulse rate values measured from the sensor can be displayed on doctors Smartphone and simultaneously stored in database. [5].

In 2017, Tigor Hamonangan Nasution, Muhammad Anggia Muchtar, Ikhsan Siregar, Ulfi Andayani, Esra Christian, worked on "Electrical Appliances Control Prototype by Using GSM Module and Arduino". Communication technology involving mobile devices and machines is growing rapidly in both industrialized and globalized world. The development of remote-control technology has grown rapidly along with the development of communication technology nowadays. GSM protocols control has also been applied to control and monitoring system with multiple devices. Here we designed a prototype to control electrical appliance via SMS using GSM SIM module 900 and Arduino. Controlling is done by a relay module via SMS controller. In addition, the controller also sends status messages from the relays. Electrical equipment settings via SMS was design in order to make the setting does not depend on specific platform of mobile devices. GSM protocols control has also been applied to control and monitoring system with multiple devices. Here we designed a prototype to control electrical appliance. The simplest communication technology available is by using GSM protocol. In this paper, a prototype of electric appliance control tool via SMS by using GSM is proposed. GSM protocol was chosen because it does not depend on mobile devices' platform. GSM SIM 900 and Arduino for controlling a relay module were utilized here. Relay module worked in accordance with orders given through SMS and the mobile device then re-

73

ceived the feedback of the command. [6].

In 2016,L. Iozzia, L. Cerina, L.T. Mainardi, worked on "Assessment of beat-to-beat heart rate detection method using a camera as contactless sensor". This paper discusses Video photo plethysmography (videoPPG) has emerged as area of great interest thanks to the possibility of remotely assessment of cardiovascular parameters, as heart rate(HR), respiration rate (RR) and heart rate variability (HRV).The present article proposes a fully automated method based on chrominance model, that selects for each subject the best region of interest(ROI) to detect and evaluate the accuracy of beat detection and interbeat intervals (IBI) measurements. the blood volume pulse signal (BVP) remotely by camera recording of human body. The physical principle is based on the study of reflected light from superficial arteries that is carrying on the pulse wave information. The possibility to record remotely video PPG signal makes this technology useful in telemonitoring, fitness and burn injuries where the use of leads is critical. Moreover, combined with facial features analysis, it could let to study emotional states of subjects during cognitive stress. [7].

In 2016, Reba Seddik ,Ayman M. Eldeib worked on "A Wireless Real Time Remote Control and Tele-Monitoring System for Mechanical Ventilators". There are many reasons lead to increase the mortality of the respiratory failure patients, who are undergoing mechanical ventilation. From these reasons, the inability of an expert physician to follow up such patients where he/she cannot stay with them all time. In this approach, we use a simple, a small, and a lightweight unit placed inside the device in an easy and a safe way. This unit allows experts to control the device remotely using a virtual keypad via internet/intranet, while monitoring the patient him/herself. Moreover, this system can tele-control the device from anywhere without causing any defect to the internal circuitry of the device. To evaluate this system, we applied it to three types of hospital mechanical ventilators with three different types of keypads. It was easy and familiar for the operator to deal with the web control page, which has the same layout of the ventilator keypad. Our system is designed for tele- monitoring a patient and tele-controlling medical devices such as hospital ventilator in real-time. Our implementation is done without making any changes in the keypad, or in the ventilator structure. Our system does not cause any influence

74

on the ventilator's internal circuitry and there is no need to modify the electrical connections of the mechanical ventilator. Once the module is connected to the device and the internet network, it is noted that the operator can select and adjust all the parameters of the ventilator in the real- time without any considerable delay.[8].

In 2013, Esuabom David Dijemeni ,Robert Dickinson worked on "Portable Mobile Real Time Oxygen Monitoring Auto Ventilation System". Respiratory diseases are pathological conditions which makes gaseous exchange harder due to the malfunction of tissue and organs responsible for breathing. A portable mobile real time oxygen monitoring auto-ventilation system using a mobile phone, oximeter, mass gas flow controller, and a portable oxygen cylinder is proposed. The system consists of a tele-monitoring system and an oxygen tele-controller system. The tele-monitoring system consists of a mobile phone and a portable oximeter. The oximeter transmits the blood oxygen level and heart rate to the mobile phone for real time monitoring. The oxygen tele- controller consists of a mobile phone, mass gas controller, oxygen cylinder, and an oxygen mask. The mobile phone is used to control the mass gas controller to supply the patient with oxygen based on the real time monitoring values. The proposed system is designed to develop a portable mobile oxygen monitoring system for oxygen delivery at home and on the go activities for Chronic Obstructive Pulmonary Disease, Obstructive Sleep Apnea, and hypoxia related disease patients. The system consists of two smaller systems, namely: monitoring unit and oxygen supply unit. The monitoring system consists of a mobile phone and an oximeter. [9].

Methodology

One of the biggest challenges in the existing system is during the time of covid or any other infectious disease, it is difficult to operate and monitor. It requires manual health checkup periodically. In the proposed system, it will control the ventilator as per the level selected. The ventilator can be accessed manually and remotely through IoT modules. The architecture of the proposed system consists of three major sections. The first section is the blood oxygen monitoring system. The second section is the oxygen control system that bridges the gap between monitoring the patient and supplying oxygen to the patient. The third section is a data transfer system for clinicians to receive the patient's monitored data. The proposed system will help to monitor patient's health continuously and updates the data in webpage. The webpage will have the patients complete health checkup data which help for future progressanalysis.

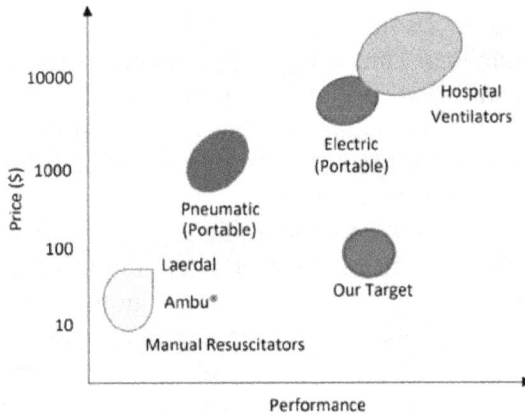

Figure 1. Expected performance level.

It is important to have affordable and high-performance systems to ensure the safety of the patients. This research deals with the design, fabrication, and characterization of one such system. The trade-offs between low production costs, ease of fabrication, and performance are considered. The combined use of the system including arduino mega and other elements are to maintain low costs without considerable compromise in performance.

Design And Implementation

This project mainly aims to make a ventilator system automatically. In this proposed system, Arduino microcontroller is used as heart of the system, here we use the high torque DC motor and keypad. The high torque DC motor is used to pump air. The keypad is used to control the three variations of the motor torque. The SPo2 and temperature sensor is used to take the patient body oxygen level and temperature level.

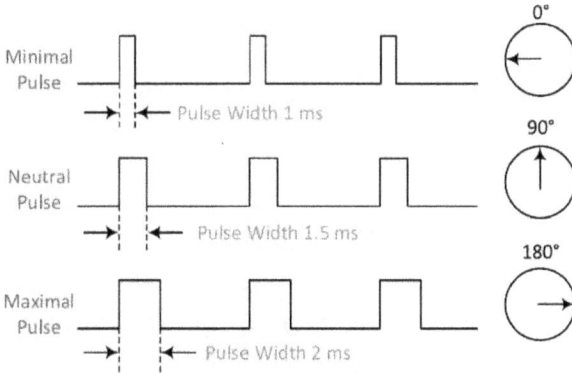

Figure 3. Servo motor pressure level.

The collected sensor values are updated with the help of the IoT module. The LCD is used to update the latest sensor values. We can control the ventilator remotely and monitor the patients' health condition through web page. The web page consist of patient complete health checkup data which helps in future progress analysis.

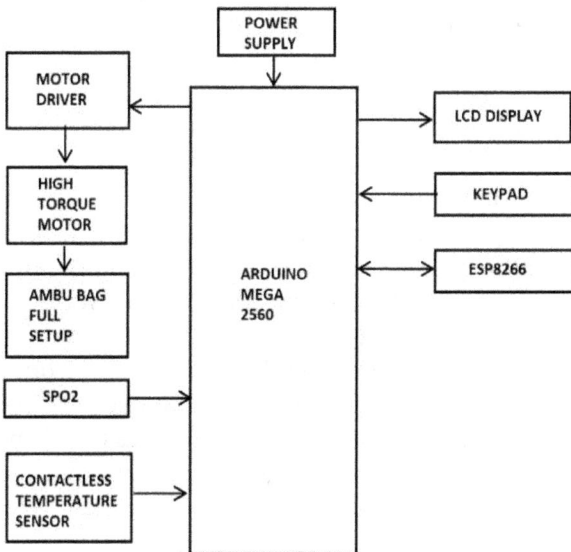

Figure 4. Block diagram of the proposed system.

Result Analysis

The portable ventilator will continuously monitor the oxygen level through the oximeter and set the motor level according to it. The temperature, oxygen level, pump level every details will be updated in the web site for remote access and future progress analysis. Our proposed system maintains the selected level in ventilator which can be controlled manually and remotely. The system continuously analyze the patient's health and updates the data in the web page. It helps the doctor to access the data from anywhere and the stored data helps in progress analysis.

Conclusion And Future Enhancement

The designing of Portable ventilator has been done using the best use of its features of Oximeter, ambu bag, IoT in order to sense and act accordingly. It includes an automated system that sends the data to thewebpage. Our proposed system maintains the selected level in ventilator which can be controlled manually and remotely. The system continuously analyze the patient's health and updates the data in the web page. It helps the doctor to access the data from anywhere and the stored data helps in progress analysis. The design was built on the main objective of cost-effectiveness and so, this system can be further improved into amore efficient portable ventilator . Day by day, inventions are being made in this field and without a doubt, they are being proven good for human safety. This technology is also designed with the hope of proving to be one among them. Future Enhancement of the project is When the bag is squeezed, the air enters the lungs of the patient, while the nonreversible breathing valve prevents backfiring of the exhaled air. Then the AMBU bag self-dispenses by sucking air from the valve from its back side. Either ambient air can be used as "fuel", or an oxygen cylinder can be connected. In the latter case, it is possible to connect a tank to collect excess oxygen, which was not used by the patient.

References

[1] Nada Y. Philip, Joel J. P. C. Rodrigues, Simon James Fong, Honggang Wang, Jia Chen,"Internet of Things for In-Home Health

Monitoring Systems: Current Advances, Challenges and Future Directions".IEEE JOURNAL ON SELECTED AREAS IN COMMUNICATIONS, VOL. 39, NO. 2, FEBRUARY 2021.

[2] Mashoedah, Umi Rochayati, Indra Hidayatulloh, Arya Sony, Ferda Ernawan Fardiansyah, Apri Nuryanto "IoT Enabled Ventilator Monitoring System for Covid-19 Patients", Journal of Physics: Conference Series, ICE-ELINVO 2021.

[3] Sivapriya Natarasan,Dr.Pavithra Sekar "Design and Implementation of Heartbeat rate and SpO2 Detector by using IoT for patients. International Conference on Electronics and Sustainable Communication Systems (ICESC 2020) IEEE.

[4] Subha R, Haritha M, Nithishna B, Monisha S G *Coma Patient Health Monitoring System Using IoT*".2020 6th International Conference on Advanced Computing & Communication Systems (ICACCS).

[5] R. Harini, B. Rama Murthy "Development of a Wireless Blood Pressure Monitoring System by Using Smartphone". *International Journal of Advanced Research in Electronics and Communication Engineering (IJARECE) Volume 6, Issue 12, December 2017.*

[6] Tigor Hamonangan Nasution, Muhammad Anggia Muchtar, Ikhsan Siregar, Ulfi Andayani, Esra Christian,"Electrical Appliances Control Prototype by Using GSM Module and Arduino"2017 4th international conference on industrial engineering and applications.

[7] L. Iozzia, L. Cerina, L.T. Mainardi, "Assessment of beat-to-beat heart rate detection method using a camera as contactless sensor".978-1-4577-0220-4/16/ ©2016 IEEE.

[8] Reba Seddik, Ayman M. Eldeib", A Wireless Real-Time Remote Control and Tele-Monitoring System for Mechanical Ventilators"978-1-5090-2987-7116/©2016 IEEE.

[9] Esuabom David Dijemeni, Robert Dickinson "Portable Mobile Real Time Oxygen Monitoring Auto-Ventilation System "IEEE International Conference on E-Health and Bioengineering - EHB 2013

Chapter 8

Privacy Preservation Methods in Big Data Analytics

Mahapatra L N N S Sri Harika, Ambati Keerthi, Chinni Sai Lakshmi
and Avula Chandra Sekhara
Sree Vidyanikethan Engineering college, Tirupathi, India

Abstract

While several research efforts are developed within the framework
of privacy-preserving big-data management and analytics recently,
relevant challenges arise when such models, techniques, and algorithms must be delivered on top of massive, distributed big data
repositories. This problem opens the door to the planning of innovative models and algorithms that, contrary to actual proposals, can
inject the scalability feature during the privacy-preserving big-data
management and analytics phase. supported these considerations,
this paper provides an outline of real-world problems and limita-
tions of state-of-the-art techniques, together with the venturing of a
requisite framework for supporting scalable privacy-preserving
big-data management and analytics.

Introduction

There is an exponential growth in volume and form of data thanks
to the varied applications of computers altogether domain areas.
This growth occurred as the affordable availability of technology,
storage, and network connectivity. the information holder can release this data to a third-party data analyst to achieve deeper in-
sights and identify hidden patterns which help make important decisions that will help in improving businesses, providing value-
added services to customers, prediction, forecasting, and recom-
mendation. Always there exists a trade-off between data utility and
privacy. This paper also proposes a knowledge lake-based modern-
istic privacy preservation technique to handle privacy preservation
in unstructured data with maximum data utility. There are three
entities liable for ensuring privacy preservation.

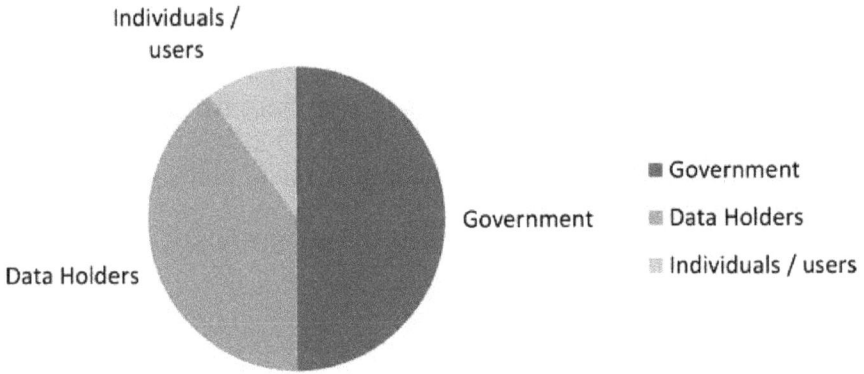

Individuals / users

Data Holders

Government

■ Government
▥ Data Holders
▧ Individuals / users

CLOUD computing and massive Data, two disruptive trends nowadays, impose a huge impact on the present industry and research community. Today, big-data services are deployed or migrated to the cloud for data processing, processing, or sharing. The salient characteristics of cloud computing like high scalability and pay-as-you-go fashion make Big Data inevitably accessible by various organizations through public cloud infrastructure.

Privacy Threats in Data Analytics

Privacy is the ability of a person to see what data to share and use access control. If the information is an exceeding property right, then it's a threat to individual privacy because it is being held by the information holder. A number of the vital privacy threats include:

S.no.	Application type	Privacy risk involved
1	Smart phone apps.	Information theft, Intrusion
2	e-Commerce sites	Inference attacks, Disclosure
3	Social media	Cyber stalking, Ransom ware
4	Data capturing systems like banking, hospitals, insurance, government portals etc.	Disclosure, Discrimination

Data analytics activity will affect data privacy. Many countries are enforcing Privacy preservation laws. Lack of awareness is additionally one of the explanations for privacy attacks. Many smartphone users don't seem to be attentive to the knowledge being stolen from their phones by many apps. Previous research shows that only 17% of smartphone users are awake to privacy threats.

Privacy Preservation Methods

MAP-REDUCE

In general, this MapReduce algorithm has split into two components: 'Map' and 'Reduce'-
1. The Map task removes data sets and converts them into another
data set, where individual data sets will be divided into key-value
pairs (or you'll call them Tuples).
2. The Reduce task will take the output data sets from the Map task
as an input value and combines them into tuples of key-value pairs.

Traditional enterprise systems accommodate a centralized server
to process data and also help to store them. Normally, traditional
models aren't suitable to process an outsized volume of information
and can't access them using standard database servers. In general, a
centralized server usually consists of the bottleneck to processing
multiple data files.

The following diagram will illustrate the bottleneck process:

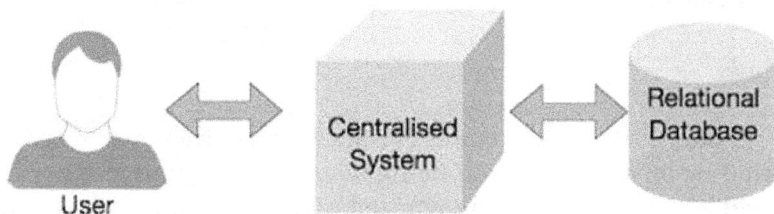

Now Google has come up with an answer to beat this bottleneck
problem popularly called the "MapReduce algorithm". Map-reduce
is an application programming model employed by big data to process data in multiple parallel nodes. Usually, this MapReduce divides a task into smaller parts and assigns them to several devices.
Then the results are collected in one place and integrated to make
effective data sets.

The below diagram will explain how this MapReduce integrates the
tasks:

83

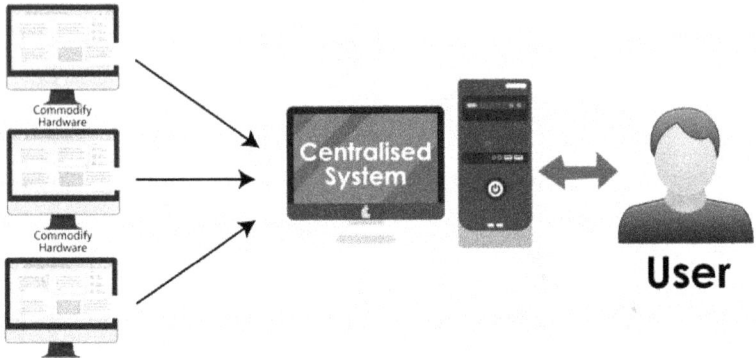

K Anonymity

k-anonymity requirement

Each release of knowledge must be such every combination of values of quasi-identifiers will be indistinctly matched to a minimum of k individuals.

Methods

Suppression: during this method, certain values of the attributes are replaced by an asterisk '*'. All or some values of a column are also replaced by '*'. within the anonymized table below, we've replaced all the values within the 'Name' attribute and every one of the values within the 'Religion' attribute with a '*'.

Generalization: during this method, individual values of attributes are replaced with a broader category. as an example, the worth '19' of the attribute 'Age' could also be replaced by ' ≤ 20', the worth '23' by '20 < Age ≤ 30', etc.,

Before anonymization

Name	Age	Gender	State of domicile	Religion	Disease
Ramsha	30	Female	Tamil Nadu	Hindu	Cancer
Yadu	24	Female	Kerala	Hindu	Viral infection
Salima	28	Female	Tamil Nadu	Muslim	TB
Sunny	27	Male	Karnataka	Parsi	No illness
Joan	24	Female	Kerala	Christian	Heart-related
Bahuksana	23	Male	Karnataka	Buddhist	TB
Rambha	19	Male	Kerala	Hindu	Cancer
Kishor	29	Male	Karnataka	Hindu	Heart-related
Johnson	17	Male	Kerala	Christian	Heart-related
John	19	Male	Kerala	Christian	Viral infection

After anonymization

Name	Age	Gender	State of domicile	Religion	Disease
*	$20 < Age \leq 30$	Female	Tamil Nadu	*	Cancer
*	$20 < Age \leq 30$	Female	Kerala	*	Viral infection
*	$20 < Age \leq 30$	Female	Tamil Nadu	*	TB
*	$20 < Age \leq 30$	Male	Karnataka	*	No illness
*	$20 < Age \leq 30$	Female	Kerala	*	Heart-related
*	$20 < Age \leq 30$	Male	Karnataka	*	TB
*	$Age \leq 20$	Male	Kerala	*	Cancer
*	$20 < Age \leq 30$	Male	Karnataka	*	Heart-related
*	$Age \leq 20$	Male	Kerala	*	Heart-related
*	$Age \leq 20$	Male	Kerala	*	Viral infection

Computing L-Diversity for a Dataset

L-diversity could be a property of a dataset and an extension of k-anonymity that measures the range of sensitive values for every column during which they occur. A dataset has l-diversity if, for each set of rows with identical quasi-identifiers, there are a minimum of l distinct values for every sensitive attribute. you'll be able to compute the l-diversity value supported by one or more columns, or fields, of a dataset.

Implementing L diversity isn't possible on every occasion due to the range of knowledge. L diversity is additionally susceptible to skewness attack. When the distribution of information is skewed into some equivalence classes attribute disclosure can't be ensured. as an example, if the complete records are distributed into only three equivalence classes, then the semantic closeness of those values may result in attribute disclosure. Also, L diversity may result in similarity attacks.

T Closeness
Another improvement to L diversity is that the T closeness measure where an equivalence class is taken into account to own 'T closeness' if the space between the distributions of a sensitive attribute within the class isn't quite a threshold and every one equivalence class has T closeness. T closeness is often calculated on every attribute concerning a sensitive attribute. T closeness may ensure attribute disclosure but implementing T closeness might not provide a proper distribution of information on every occasion.

Conclusion

Big-data privacy has become an important issue since it is directly related to customers. It is now essential for an organization to promise privacy in big data analytics. Privacy measures should now focus on the uses of data rather than the collection of data. They should be modified concerning the size and unexpected uses of big data. Techniques like anonymization have limited potential when applied to big data.

References

B.C.M. Fung, K. Wang, R. Chen and P.S. Yu, "Privacy-Preserving Data Publishing: A Survey of RecentDevelopments," ACM Comput. Surv., vol. 42, no. 4, pp. 1-53, 2010.

K. Lefevre, D.J. DeWitt and R. Ramakrishnan, "Workload- Aware Anonymization Techniques for Large-Scale Datasets," ACM Trans. Database Syst., vol. 33, no. 3, pp. 1-47, 2008.

Andrea C. Arpaci-Dusseau, Remzi H. Arpaci-Dusseau, David E. Culler, Joseph M. Hellerstein, and David A. Patterson. High-performance sorting on networks of work-stations. In Proceedings of the 1997 ACM SIGMOD International Conference on Management of Data, Tucson, Arizona, May 1997.

Remzi H. Arpaci-Dusseau, Eric Anderson, Noah Treuhaft, David E. Culler, Joseph M. Hellerstein, David Patterson, and Kathy Yelick. Cluster I/O with River: Making the fast case common. In Proceedings of the Sixth Workshop on Input/Output in Parallel and Distributed Systems (IOPADS '99), pages 10–22, Atlanta, Georgia, May 1999.

Chapter 9

Manhole Monitoring System

V.M.Karthikeyan, Balaganesh R. and Jayaseelan S.
Department of Electronics and Communication Engineering, St. Joseph's Institute of Technology, Chennai, India.

Abstract

A smart city is the future goal to have cleaner and better amenities for the society. Smart underground infrastructure is an important featureto be considered while implementing a smart city. Drainage system monitoring plays a vital role in keeping the city clean and healthy. Since manual monitoring is incompetent, this leads to slow handling of problems in drainage and consumes more time to solve. To mitigate all these issues, the system using a wireless sensor network, consisting of sensor nodes is designed. The proposed system islow cost, low maintenance, IoT based real timewhich alerts the managing station through messagewhen any manhole crosses its threshold values. This system reduces the death risk of manual scavengers who clean the underground drainage and also benefits the public.

Introduction

An integral part of any drainage system is theaccess points into it when it comes to cleaning, clearing, and inspection. Metropolitan cities have adopted underground drainage system and the city's municipal corporation must maintain its cleanliness. If the sewage maintenance is not proper, ground water gets contaminated causing infectious diseases. Blockages in drains during monsoon season, causes problems in the routine of the public. Hence, there should be a facility in the city's corporation, which alerts the officials about blockages in sewers, their exact location. It mainly acknowledges in the field of alerting the people about the gas explosion, increase in the water leveland the temperature level. It uses IoT to make the drainage monitoring system in a highly automotive by using sensor for detecting and sending alerts GSM module to the authorities. This project overcomes the demerits by detectingdrainage water blockage by installing water flow ratesensors at the intersection of nodes. When there is a blockage in a particular node, there is variation in the flow of drainage water which when cross the set value will display the alert in the managing station. Other demerits are solved by detecting temperature variations inside the manhole and alerting the same to the managing station. Also, flowrate sensors are used to detect the overflow of the drainage water and alerting the same to the managing station through automatic message. Maintenance of manholes manually is tedious and dangerous due to the poor environmental conditions inside so, the mainfocus of this project is to provide a system which monitors water level, atmospheric temperature, waterflow and toxic gases.

If drainage gets blocked and sewage water overflows, it is sensed by the sensors and message is sent to the municipal. It is, therefore, dangerous to goinside the manholes for inspection of its current state.To solve all the problems related to undergroundsanitation, a remote alarm system is necessary for transmitting data collected by the sensors set inside the manhole to the managing station. This includes components such as controller, memory, transceiver and battery to supply power.

Literature Review

In 2020 He-sheng Zhang; Lei Li; Xuan Liu worked on "Development and Test of Manhole Cover Monitoring Device Using LoRa and Accelerometer".The monitoring device is installed under the manhole cover, three key problems, how to measure the tilt angle or state of resources °. For the second problem, long range (LoRa) is adopted. A 433-MHz whip antenna is designed to overcome the shield of manhole cover and the absorption of electromagnetic waves of the earth. The field tests show that the effective communication distance has been extended more than 700 m by using the whip antenna. In addition, the parameter spreading factor (SF) and bandwidth (BW) are configured. For the third problem, sleep mode is used, and input output (IO) pin is configured.

In 2018 W.Z.Black; Robert E.Snodgrass; Bryan P.Walsh worked on Thermodynamic Analysis of Explosive Events in Manholes and Electrical Vaults. A computer code is used to simulate theconditions that exist during a manhole event consisting of an electrical fault and a gas explosion. When energy is generated within the structure, pressures can reach levels sufficient to propel the cover from its frame. The code quantifies the effects of the event and provides guidelines for the design of safety devices that can minimize the potential danger of the event. Differences between chemically driven and electrically-driven events are shown to require different design criteria for cover restraint systems. Devices designed to restrain the cover must be able to withstand explosive forces and react in sufficient time to reduce internally generated pressures.

In 2019 Glen Bertini worked on "Manhole explosion and its root causes". The phrase "manhole events" is a euphemism for fires and explosions that occur in utility manholes in urban areas. All fires and explosions require three elements: fuel, oxygen (or another oxidizer), and an ignition source. Further, the fuel and oxygen must be within aspecified range of concentrations to support combustion. Figure1 lists most of the significant compounds (flammable or not) that are likely to be encountered in a manhole environment and, where applicable, their upper and lower explosive limits (UELs and LELs, respectively).

In 2018, N Nataraja; R Amruthavarshini; N L Chaitra; K Jyothi; N Krupaa; S S M Saqquaf worked on "Secure Manhole Monitoring system Employing sensor and GSM Techniques". Nowadays manhole problems in the populated cities are the major issues. Opening of manholes due to breakage of manhole cover, manhole explosions are major threat in recent days. Manhole cover opening leads to accidental fall ofvehicles, pedestrians leading to accidents or loss of life. Manhole opening detection and alerting ismainly based on detecting the manholes which areopened due to overflow of sewage / rainwater during heavy rainfall and alerting When amanhole opening is detected either due to overflow of sewage water, increase in pressure or temperature, it leads to the breakage of the manhole lids. To avoid such incidents even beforeit could affect the public, an alerting system is built wherein the buzzer alerts the surrounding and sends the sensed data to the managing authorities using GSM techniques.So, they can take precautionary action to close the manhole considering public safety.

In 2016, Sugato Ghosh; Indranil Das; Deepanjana Adak; Nillohit Mukherjee; Raghunath Bhattacharyya worked on "Development of selective and sensitive gas sensors for manhole gas detection" worked on Loss of life of the workers inside manhole is a common problem for many parts of the world. To resolve this issue a portable, low cost,simple manhole gas detection unit has been designed and developed which is capable to detect the poisonous gases like carbon monoxide, hydrogen sulfide, and explosive gaslike methane within a minute and raise alarm if the concentration of any gas is beyond the threshold limit.

Exisiting Method

Detecting a struck in the manhole pipe, due to blockage of sludge in a pipe. It is also necessary to alert the concerned authorities in case of opened/displaced/ breakage of manhole lids for necessary and immediate action. This enables to maintain a safe and user-friendly environment. To monitor water level in manhole sensors are to be deployed which will detect and transmit necessary data for immediate attention to clear the clogs well before it overflows.

Disadvantages Of Existing System

In this system, the Open manhole cause flooding inthe area. Flooding itself may displace populations and lead to further health problems. This means high labour cost and time consumption. Many times, the data is not maintained properly due to manual entering of the data in the registers or systems. Poor   drainage   often   occurs when contractors remove topsoil during construction of new homes, leaving only subsoil. Chance of mixing of groundwater and drainage water .if not taken proper care.

Methodology Of Proposed System

A new automated system is presented to solve the aforementioned shortcomings. The manhole is monitored by sensors in the proposed system, and datais automatically updated on a remote server. The microcontroller Arduino Mega interfaced with the several types of sensors (flow, level, temperature, andgas sensors). When the various sensors hit the threshold level, the microcontroller receives an indicator of that value and sensor is being sent to the microcontroller. Furthermore, Arduino Mega then sends the signal and location of the manhole to the municipal corporation through Alert and the officialscould easily locate which manhole is having the problem and could take appropriate steps. Also, Arduino Mega updates the live values of all the sensors. We have included an array of sensors for complete monitoring of the manhole cover so that such accidents can be prevented. This project includes a gas cover to monitor the gas emitted from the sewage systems so that toxicity can be monitored, the internal temperature is also monitored if a check for a change in the temperature as the property of manhole change with temperature which could need to crack formation, a tilt sensor is introduced to indicate whether the manhole can tilt.

GPS interface with microcontroller

The system is equipped with GPS. This GPS is a made-up location that can hold data in the form ofa link. The information to be added to the link is the location of the plugged manhole. The result isthen accessed via a web server on a PC/Laptop. Along with the ardunio mega, which is placed in a manhole, the GPS is bannered. The location iscommunicated to the web server and can beviewed there.

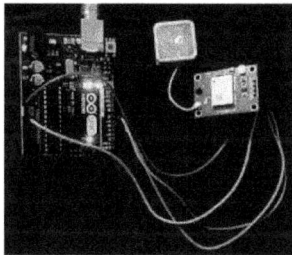

GSM interface with microcontroller

The system is equipped with GSM module. There aredifferent kinds of GSM modules available on themarket. We are using the most popular module based on Simcom SIM900 and Arduino Uno for this tutorial. Interfacing a GSM module to Arduino is pretty simple. You only need to make 3 connections between the gsm module and Arduino. the first method, you have to connect the Tx pin of the GSM module to theRx pin of Arduino and the Rx pin of the GSM moduleto the Tx pin of Arduino. GSM Tx->Arduino Rx and GSM Rx->Arduino Tx. Now connect the ground pin of Arduino to the ground pin of the gsm module.

Connect 12V 2A Adapter to the GSM

GND

Tx-->RX
Rx-->TX

Block Diagram of the Proposed System

POWER SUPPLY

ULTRASONIC SENSOR		GPS
GAS SENSOR MQ5		FLOW SENSOR
DHT11	ARDUINO MEGA	CLOUD
IR SENSOR		GSM

ALGORITHM

Start

Power the hardware components

Initialize the hardware component

IR Sensor | GAS Sensor | Level sensor | DHT11 sensor

Microcontroller takes input continuously from sensor

If any sensor exceeds its set value?

No

Yes

Location traced by using GPS module

All sensor values are displayed

Message will sent to municipal office by using GSM

Normal flow

End

94

Results

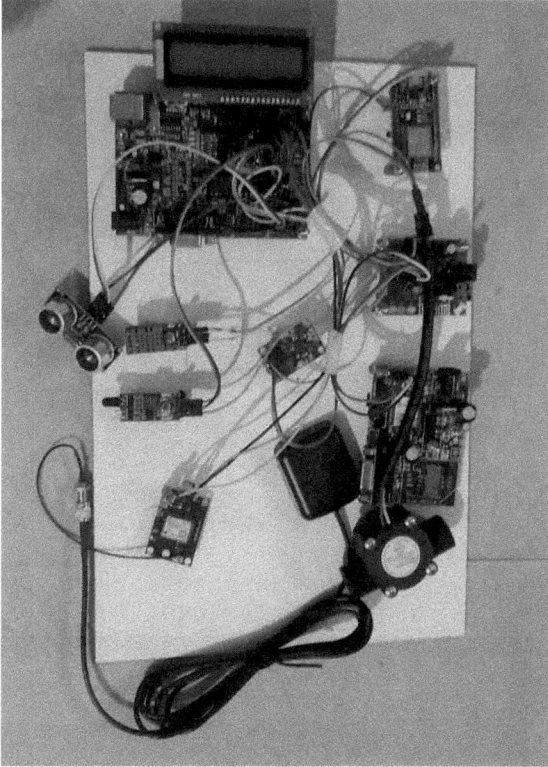

Figure 1. The setup of proposed system

The prototype of smart drainage system the sensors sense the information and report that information to Arduino Mega. The gas sensor senses the harmful gases; water flow rate sensor will check the flow of water and also check the level of water and sends the values to the Arduino Mega. It designed with inbuilt wifi module will process the information sent by the all the sensors and store that information in the cloud. Also that value will be displayed in web server. The recorded values will be displayed in 16X2 LCD. If sensed values exceed the threshold values then sends information to local municipal officer and directly to the concern area workervia SMS .The worker directly can check the status of Drainage system in their smart phone using web page and also via SMS Alert. they can take further action based on the status and also they can identify the location.

Figure 2. LCD display of drainage block.

Figure 3. LCD display.

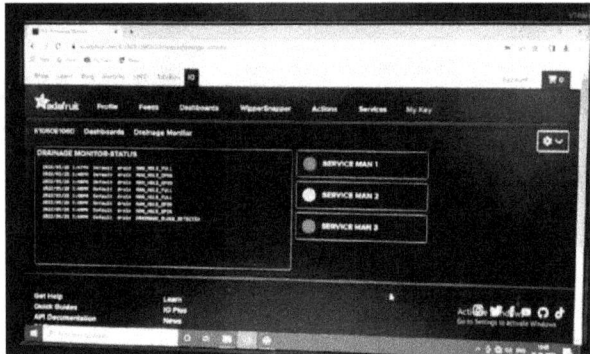

Figure 4. Web page alerts.

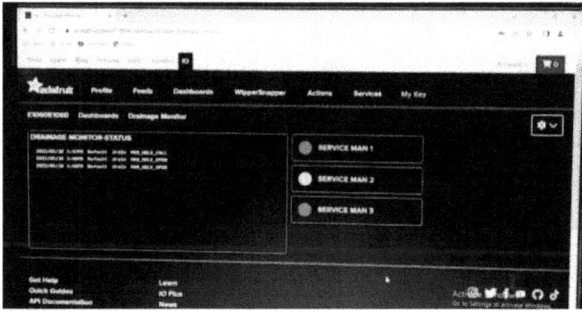

Figure 5. Web page alerts.

Figure 6. SMS alert of manhole filled

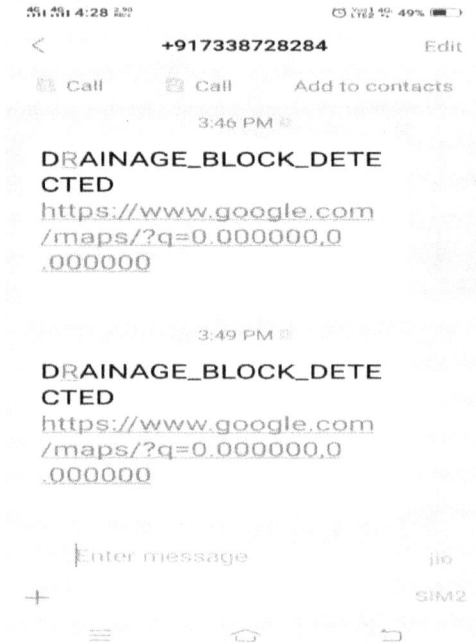

Figure 7. SMS alert of manhole block detected

Conclusion

This project proposes different methods for monitoring and managing underground drainage system. It explains various applications like underground drainage and manhole identification in real time. Various parameters like temperature, toxic gases, flow and level of water are being monitored and updated on the internet using the Internet of Things. This enables the person in-charge to take the necessary actions regarding the same. In this way the unnecessary trips on the manholes are saved and can only be conducted as and when required. Also, real time update on the internet helps in maintaining the regularity in drainage check thus avoid the hazards. Our project helps to reduce the problem of drainage system with the help of sensors like water level detection sensor and gas sensor our mechanism helps to notify the connected network, when the harmful gases are detected to gas sensor. By this project the underground drainage system can be easily organized

Future Scope

A new method combined with multi-view matching and feature extraction technique is developed to detect manhole covers on the streets using close-range images combined with IMU and LINDAR data. The covers are an important target on the road traffic as same as transport signs, traffic lights and zebra crossing but with more unified shapes. However, the different shoot angle and distance, ground material, complex street scene especially its shadow, and cars in the road have a great impact on the cover detection rate. The paper introduces a new method in edge detection and feature extraction in order to overcome these difficulties and greatly improve the detection rate.

References

o G. Bertini, "Manhole explosion and its root causes," in IEEE Electrical Insula- tion Magazine, vol. 35, no. 1, pp. 45-54, January-February 2019.

o S. Ghosh, I. Das, D. Adak, N. Mukherjee, R. Bhattacharyya and H. Saha, "De- velopment of selective and sensitive gas sensors for manhole gas detection," 2016 10th International Conference on Sensing Tech- nology (ICST), 2016.

o W. Z. Black, R. E. Snodgrass and B. P. Walsh, "Thermo- dynamic Analysis of Explosive Events in Manholes and Electrical Vaults," in IEEE Transactions on Power Deliv- ery, vol. 37, no. 1, pp. 290-297, Feb. 2022.

o W. Z. Black, R. E. Snodgrass and B. P. Walsh, "Thermo- dynamic Analysis of Explosive Events in Manholes and Electrical Vaults," in IEEE Transactions on Power Deliv- ery, vol. 37, no. 1, pp. 290-297, Feb. 2022.

o H. Zhang, L. Li and X. Liu, "Development and Test of Man- hole Cover Monitor- ing Device Using LoRa and Accel- erometer," in IEEE Transactions on Instru- mentation and Measurement, vol. 69, no. 5, pp. 2570-2580, May

2020.

o Y. Yu, J. Li, H. Guan, C. Wang and J. Yu, "Automated Detection of Road Man-hole and Sewer Well Covers From Mobile LiDAR Point Clouds," in IEEE Geo- science and Remote Sensing Letters, vol. 11, no. 9, pp. 1549-1553, Sept. 2014.

o L. Lei, Z. H. Sheng and L. Xuan, "Development of low power consumption man - hole cover monitoring device using LoRa," 2019 IEEE International Instrumentation and Measurement Technology Conference (I2MTC), 2019.

Chapter 10

Implementation of Trust Aware Random Forest and Extreme Gradient Boosting Algorithm for Safe and Secured Communication in VANET

S. Ganesh and S. Kannadhasan
Department of Computer Science and Engineering, Study World College of Engineering, Coimbatore, India

Abstract

Vehicle Ad Hoc Networking, or VANET, is a wireless communication method. This method is used to increase traffic efficiency and communication between cars. The most important aspect of the system is the communication between V2v and V2i, hence appropriate security and privacy measures must be maintained. We suggest an authentication method based on the hash function to increase the security and effectiveness of the VANET and to stop cars from interfering with one another illegally. The Chinese remainder theorem is used in the group key agreement system, and only an authorized vehicle may utilize it. The forward and backward secrecy has been accomplished in this manner. Our plan fulfils security and privacy requirements.

Introduction

The increasing number of vehicles in the city, Intelligent Transportation System (ITS) is one of the most promising directions for the efficient management of urban transport. Based on the concept of Mobile Ad-hoc Network (MANET), researchers and vehicle manufacturers have introduced the Vehicular Ad-hoc Networks (VANETs) to build the next-generation transportation systems. As a variant of MANET, VANET is a distributed, less infrastructure, self-organizing communication network, which is built among moving vehicles.

VANET aims to provide attractive services such as security services, including curve speed warnings, emergency vehicle warnings, lane changing assistance, pedestrian crossing warnings, traffic violation warnings, road intersection warnings, and road condition warnings. Moreover, it can also provide weather information, traffic information, gas station or restaurant location and other comfort and interactive services such as Inter-net access. In general, VANET is mainly composed of three parts:

1. Trusted authority (TA)
2. Roadside unit (RSU)
3. Vehicle

The TA provides the necessary network connection and stores information about the vehicle and the RSU. Each vehicle is equipped with an on-board unit (OBU), which is a tamper-proof device. Vehicles exchange traffic information through wireless communication. There are two types of communication in VANETs: vehicle-to-vehicle (V2V) and vehicle-to-infrastructure (V2I).

In V2V communication, vehicles in the same RSU range are allowed to exchange secret information. Meanwhile, V2I communication is mainly between the vehicle and an RSU fixed on the side of the road.

Authentication

The process of identifying an individual usually based on a username and password. In security systems, authentication is distinct from authorization, which is the process of giving individuals access to system objects based on their identity.

Purpose Of Authentication

- To control access to a system or application
- To bind some sensitive data to an individual, such as for encryption
- To establish trust between multiple parties to form some interaction with them.

Vehicular Ad-Hoc Networks (VANET)

VANET means vehicular Ad hoc network and it is the technology which is used to move vehicles as joint in network to make a transportable network. Participating vehicles become a wireless connection or router through VANET and it allow the vehicles almost to connect 100 to 300 meters to each other and in order to create a wide range network, other vehicles are connected to each other so the mobile internet is made. It is supposed that the first networks that will incorporate this technology are fire and police mobiles to interact with one another for security reasons.

Technology Used

Brilliant way to use Vehicular Networking is defined in VANET or Intelligent Vehicular Ad-Hoc Networking. Multiple ad-hoc networking technologies integrated in VANET such as, ZigBee, IRA, WiMAX IEEE, and Wi-Fi IEEE for convenient, effective, exact, simple and plain communication within automobiles on active mobility. Useful procedures like communication of media within automobiles can be allowed process as well to follow the automotive automobiles are also favored.

Security measures are defined in vehicles by VANET, flowing communication within automobiles, edutainment and telemetric. Selection of wireless technologies are required to implement in VANET as DSRC (Dedicated Short Range Communication) which is include in Wi-Fi. Other entrant technologies of wireless are Satellite, Wi-MAX, and Cellular.

Vehicular Ad-hoc Networks (VANET) can be considered as device of ITS (Intelligent Transportation Systems). ITS (Intelligent Transportation Systems) has conceived vehicular networks. IVC (Inter-Vehicle Communication) permits the automobile communicate to each other at the same time RVC (Roadside-to-Vehicle Communication) allows with the stations-based wayside.

Scope

The most favorable target is the more useful, efficient and safer roads will build through vehicular networks by informing to basic authorities and drivers in time in the future. Another target is to discover the advancement of vehicular ad hoc networking (VANET) wireless technologies.

The purpose is to secure and to make possible commercial requests through range of communication systems and/or other networks (VANET) which goes short to medium. These technologies would support main concern for critical time secure communication and fulfill the QOS needs of other multimedia software or e-commerce mobile.

Next goal to create high-presentation, extremely measurable and secured technologies of VANET shows an unusual challenge to the investigate community of wireless. Specific restrictions normally assumed in ad hoc networks are alleviated in VANET yet. Such as, VANET might assemble comparatively huge means of computational.

Features of VANET

Mostly interests to MANETS belong to the VANETS but the features are different. Vehicles are likely to move in structured way. The connection with wayside equipment can similarly be indicated absolutely accurately. In the end, mostly automobiles are limited in their motion range, such as being controlled to pursue a paved way.

VANET suggests unlimited advantage to companies of any size. Vehicles access of fast speed internet which will change the automobiles' on-board system from an effective widget to necessary productivity equipment, making nearly any internet technology accessible in the vehicle. Thus, this network does pretend specific security concerns as one problem is no one can type an email during driving safely.

This is not a potential limit of VANET as productivity equipment. It permits the time which has wasted for something in waiting called "dead time", has turned into the time which is used to achieve tasks called "live time".

If a traveler downloads his email, he can transform jam traffic into a productive task and read on-board system and read it himself if traffic stuck. One can browse the internet when someone is waiting in vehicle for a relative or friend.

If GPS system is integrated, it can give us a benefit about traffic related to reports to support the fastest way to work. Finally, it would permit for free, like Skype or Google Talk services within workers, reducing telecommunications charges.

The following is a list of some ad hoc network routing protocols.

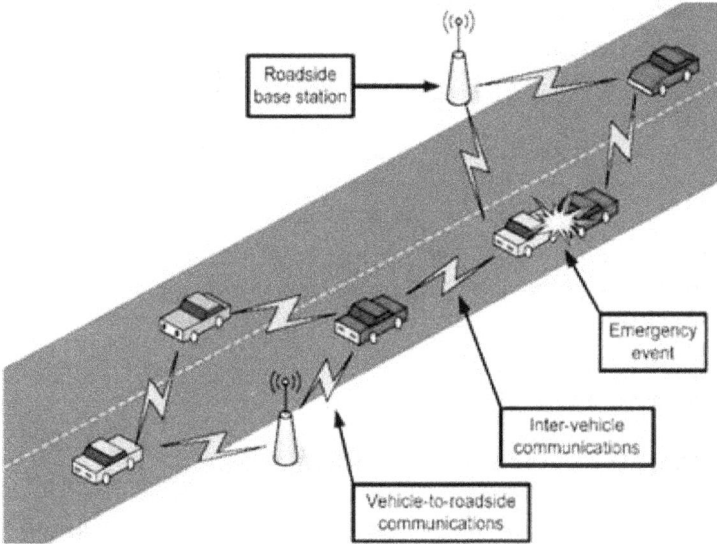

Figure 1. Avoiding road accident.

Figure 2. Traffic monitoring.

Routing Protocols

An ad-hoc routing protocol is a convention, or standard, that controls how nodes decide which way to route packets between computing devices in a mobile ad hoc network.

In ad-hoc networks, nodes are not familiar with the topology of their networks. Instead, they have to discover it. The basic idea is that a new node may announce its presence and should listen for announcements broadcast by its neighbors. Each node learns about nodes nearby and how to reach them, and may announce that it, too, can reach them.

Note that in a wider sense, ad hoc protocol can also be used literally, that is, to mean an improvised and often impromptu protocol established for a specific purpose.

Topology based Routing Protocols

Table-driven (Proactive) Routing Protocols

The proactive routing means that the routing information like next forwarding hope is maintained in the

Background irrespective of communication requests. The packets are constantly broadcast and flooded among nodes to maintain the path, then a table is constructed within a node which indicates next hop node towards a destination.

The advantage of proactive routing protocols is that there is no route discovery is required since the destination route is stored in the background, but the disadvantage of this protocol is that it provides low latency for real time application, it also leads to the maintenance of unused data paths, which causes the reduction in the available bandwidth. The various types of proactive routing protocols are:

(a) Fisheye State Routing (FSR)

FSR is similar to LSR, in FSR node maintains a topology table (TT) based upon the latest information received from neighboring and periodically exchange it with local neighbors. For large networks to reduce the size of message the FSR uses the different exchange period for different entries in routing tables.

Routing table entries for a given destination are updated preferably with the neighbors having low frequency, as the distance to destination increases. The problem with the FSR routing is that with the

As the mobility increases route to remote destination become less accurate. If the target node lies out of scope of source node then route discovery fails.

(b) Temporally Ordered Routing Algorithm (TORA)

TORA belongs to the family of link reversal routing in which directed a cyclic graph is built which directs the flow of packets and ensures its reach ability to all nodes. A node would construct the directed graph by broadcasting query packets. On receiving a query packet, if node has a downward link to destination it will broadcast a reply packet; otherwise it simply drops the packet.

A node on receiving a reply packet will update its height only if the height of replied packet is minimum of other all the reply packets. TORA Algorithm has the advantage that it gives a route to nodes in the network, but the maintenance of all these routes is difficult in VANET.

On Demand (Reactive) Routing

This type of protocols finds a route on demand by flooding the network with Route Request packets. The main disadvantages of such algorithms are:
- High latency time in route finding.
- Excessive flooding can lead to network clogging.

(A) Ad hoc on demand distance vector routing

Ad hoc on demand distance vector routing is an improvement on the DSDV algorithm AODV minimizes the number of broadcast by creating routes on demand as opposed to DSDV that maintains the list of all the routers.

To find a part to the destination the source broadcast a route request packet. The neighbor's intern broadcast the packet to their neighbors till it reaches an intermediate node that has recent route information about the destination or till it reaches the destination.

(B) Cluster based protocol

In cluster-based routing protocol the nodes are divided into clusters. When a source has to send data to destination it floods route request packets on receiving the request a cluster head checks to see if the destination is in its cluster in CVRP routing is done using source routing.

Position Based Routing Protocol

Position based routing consists of class of routing algorithm. They share the property of using geographic positioning information in order to select the next forwarding hops. The packet is sent without any map knowledge to the one hop neighbor which is closest to destination.

Position based routing is beneficial since no global route from source node to destination node need to be created and maintained.

Position based routing is broadly divided in two types: Position based greedy V2V protocols, Delay Tolerant Protocols.

Position Based Greedy V2V Protocols

In greedy strategy and intermediate node in the route forward message to the farthest neighbor in the direction of the next destination. Greedy approach requires that intermediate node should possessed position of itself, position of its neighbor and destination position.

The goal of these protocols is to transmit data packets to destination as soon as possible that is why these are also known as min delay routing protocols. Various types of position based greedy V2V protocols are GSR, GPSR, SAR, GPCR, CAR, ASTAR, STBR, CBF, DIR and ROMSGP.

Geographic Source Routing (GSR)

Earlier GSR was used in MANET. Then it was improved to use in VANET scenario by incorporating in to it greedy forwarding of messages toward the destination. If at any hop there are no nodes in the direction of destination, then GPSR utilizes a recovery strategy known as perimeter mode. The perimeter mode has two components one is distributed planarization algorithm that makes local conversion of connectivity graph into planar graph by removing redundant edges.

Second component is online routing algorithm that operates on planer graphs. So, in VANET perimeter mode of GPSR is used.

In GPSR if any obstruction or void occurs then algorithm enter perimeter mode and planner graph routing algorithm start operations, it involves sending the message to intermediate neighbor instead of sending to farthest node, but this method introduces long delays due to greater no. of hop counts.

Due to rapid movement of vehicles, routing loops are introduced which causes dissemination of messages to long path. GPSR uses static street map and location information about each node, since GPSR does not consider vehicle density of streets, so it is not an efficient method for VANET.

Hybrid Protocols

The hybrid protocols are introduced to reduce the control overhead of proactive routing protocols and decrease the initial route discovery delay in reactive routing protocols.

ZRP: Zone routing protocol

In this the network is divided into overlapping zones. The zone is defined as a collection of nodes which are in a zone radius. The size of a zone is determined by a radius of length α where α is the number of hops to the perimeter of the zone.

In ZRP, a proactive routing protocol (IARP) is used in intra-zone communication and an inner-zone reactive routing protocol (IARP)

109

is used in intra-zone communication. Source sends data directly to the destination if both are in same routing zone otherwise IERP reactively initiates a route discovery.

HARP: Hybrid Ad Hoc Routing Protocol

It divides entire network into non-overlapping zones. It aims to establish a stable route from a source to a destination to improve delay. It applies route discovery between zones to limit flooding in the network, and choose best route based on the stability criteria. In HARP routing is performed on two levels: intra-zone and inter-zone, depending on the position of destination. It uses proactive and reactive protocols in intra zone and inter-zone routing respectively. It is not applicable in high mobility ad hoc networks.

Security Vulnerabilities of VANETS

Vehicular ad hoc networks are also prone to several vulnerabilities and attacks. These vulnerabilities can cause small to severe problems in the network and also poses some potential security threats which can deteriorate their functioning.

The following section gives a general overview of Vehicular Communications vulnerabilities.

Jamming: The jammer deliberately generates interfering transmissions that prevent communication within their reception range. Fig. 1 illustrates that an attacker can relatively easily partition the vehicular network. As the network coverage area (e.g., along a highway) can be well-defined, at least locally, jamming is a low effort exploit opportunity.

Forgery: The correctness and timely receipt of application data is major vulnerability. The attacker forges and transmits false hazard warnings which are taken up by all vehicles.

Impersonation: Message fabrication, alteration, and replay can also be used towards impersonation. For example, an attacker can masquerade as an emergency vehicle to mislead other vehicles to slow down and yield. A vehicle owner deliberately stealing another vehicle's identity and attributing it to his or her own car or vice versa.

Privacy: The inferences on driver's personal data could be made, and thus violating his or her privacy. The vulnerability lies in the periodic and frequent vehicular network traffic: Safety and traffic management messages, transaction-based communications (e.g., automated payments).

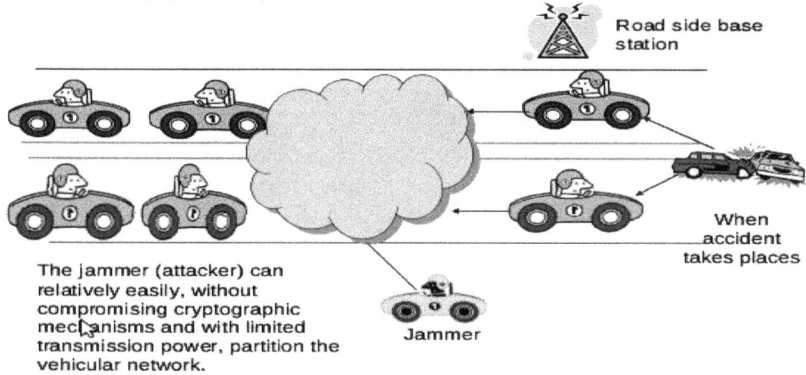

Figure 3. Jamming.

False Position Information

In VANETs, one critical issue is that when nodes send false position information in their beacon messages, which can severely impact the performance of the network. A potential source for such false position data is malicious nodes. Hence Security in VANETs relies upon the potentially more challenging problem of detecting and correcting malicious data.

VANETs have special requirements in terms of node mobility and position-dependent applications, which are well met by geographic routing protocols. One critical issue is that when nodes send false position information in their beacon messages, this can severely impact the performance of the network.

A potential source for such false position data is malicious nodes. The intents of an adversary may range from simply disturbing the proper operation of the system to intercepting traffic exchanged by ordinary users, followed by a potential modification and retransmission.

This section outlines the effects presented in which are caused by falsified position information. Fig. 2 shows an example scenario where node A claims to be at two additional (faked) positions Avi and Avr. Based on a greedy forwarding strategy nodes always select the node nearest to the destination as the next forwarding node. Assuming that F wants to send a packet to node K, it will first send the packet to its only direct neighbor G.

G will then forward the packet to the node nearest to the destination from which it received beacons. This seems to be Avr, so the packet ends up at node A, which can now forward, modify, or discard it at will. In the opposite direction, the packet from K will go to I, which will again send it to the assumed best node Avi.

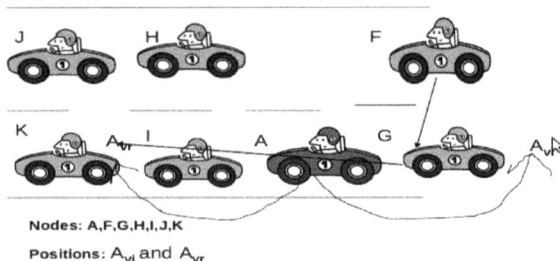

Nodes: A,F,G,H,I,J,K
Positions: A_{vi} and A_{vr}

Figure 4. Falsified position information.

Applications for VANETS

Public Safety Applications
Traffic Management Applications
Traffic Coordination and Assistance Applications
Traveller Information Support Applications
Comfort Applications
Air pollution emission measurement and reduction
Law enforcement

Public Safety Applications

- Applications are geared primarily toward avoiding accidents and loss of life of the occupants of vehicles.
- Collision warning systems have the potential to reduce the number of vehicle collisions in several scenarios.
- Safety applications have obvious real-time constraints, as drivers have to be notified before the information is no longer useful. Either an MIVC or a URVC (SRVC for intersections) can be used for these applications. It is possible that, depending on the communication range, an SIVC may be sufficient for these applications.
- In terms of addressing, the destinations in these applications will not be individual vehicles, but rather any relevant vehicle. The zone of relevance (ZOR) (also known as the target area) is determined by the particular application.

Traffic Management Applications

- Traffic management applications are focused on improving traffic flow, thus reducing both congestion as well as accidents resulting from congestion, and reducing travel time
- Traffic monitoring
- Traffic light scheduling
- Emergency vehicles

Traffic Coordination and Assistance Applications

- Platoon (i.e., forming tight columns of vehicles closely following each other on highways)
- Passing and lane change assistance may reduce or eliminate risks during these maneuvers, since they are often the source of serious accidents.

Traveler Information Support Applications

Local information such as local updated maps, the location of gas stations, parking areas, and schedules of local museums can be downloaded from selected infrastructure places or from other "local" vehicles. Advertisements with, for example, gas or hamburger prices may be sent to approaching vehicles.

Road warnings of many types (e.g., ice, oil, or water on the road, low bridges, or bumps) may easily be deployed by authorities simply by dropping a beacon in the relevant area.

Comfort Applications

This class of applications may be motivated by the desire of passengers to communicate with either other vehicles or ground-based destinations such as Internet hosts or the public service telephone network (PSTN).

Voice, instant messaging, or similar communications may occur between the occupants of vehicle caravans traveling together for long distances, or between law enforcement vehicles and their "victims."

Note that this application does not scale to large network sizes.

Vehicle to land-based destination communications is arguably a very useful capability as it may enable an entire array of applications, from email and media streaming to Web browsing and voice over IP.

Unfortunately, land-based access *requires a URVC* system that may be prohibitively expensive in the near future.

Tolls for roads and bridges can be collected automatically. Many nonstandard systems exist and work well.

Parking payments can be made promptly and conveniently.

Repair and maintenance records can be recorded at the garages performing them.

Multimedia files such as DVDs, music, news, audio books, pre-recorded shows can be uploaded to the car's entertainment system while the car is in the garage.

Existing System

Destination zones and segments lower dynamic and higher dynamic VANETs scenarios were controlled. Based on road segmentation technique. Failed to consider dynamically adjusted. **Distributed version of Time-Limited Reliable Broadcast Incremental Power (DTRBIP) algorithm** was introduced in [1]

The main objective of DTRBIP algorithm was, to manage both delay-sensitive and delay-tolerant applications in urban and highway environment. However, DTRBIP algorithm sizes.

In order to perform authentication of routing messages in VANET, **Trustworthy VANET Routing with group authentication keys (TROPHY) protocol** was developed in [2] with WAVE architecture and patented routing approach For refreshing the cryptographic material and preserving the authentication keys updated across network authorized nodes were employed to gather TROPHY messages However, the temporary cooperation failed to provide without compromising any companies

Based on message authentication, **Elliptic Curve Digital Signature Algorithm (ECDSA)** was designed in [3] for multi-hop routing control plane in VANET. In order to maintain the distributed routing path control plane employed to update periodic locality. The multiple curve parameters linked with WAVE and ETSI ITS. Through different cryptographic libraries. However, during fast electric charging of electric vehicles. ECDSA validations were not used for performance evaluation.

CPS presented defends against syntactic linking attack, semantic linking attack, misleading attack/false alarm, Sybil attack and impersonation attack. The congestion and communication overhead were reduced, and QoS maintained with location protection.

A **dynamic entity-centric trust model** depending on weight was developed in [7] consistent with applications and nodes authority level Security level of routing protocol GPSR was enhanced using trust model with minimum delay and maximum success data delivery rate. To realize the trust evaluation in VANET simple data-centric trust model was designed But, computational complexity was not reduced without compromising delay

Disadvantages

- Low efficiency

- Complexes the process

Proposed Work

The algorithm proposed finds a perfect route to a source and sends the message via the route it has discovered. It uses the multi-objective bio-inspired heuristic cuckoo search optimization algorithm to find out the optimum route. The steps for finding the best route as depicted.

System model

Based on the following assumption, the VANET model is used to select the best routes between the source and its destinations. The routing in the network is the special type of network that provides its customers with a variety of services, such as message alerts, driver predictions, and information services.

In many ITS-related applications, such as brake warning, weather forecasting, traffic, vehicle safety, the road network has been of the very important role. There is, therefore, a major requirement for a communication system between vehicles which is reliable, stable and entitled to provide a security message in due course. There have been several attempts to standardize road network design and deployment.

The network model to be considered for the establishment of the road model structure. Let N be the average number of network vehicle comprising a number of segments will be figure. Here out route discovery process provides the origin with knowledge about the destination path.

Proposed Algorithm

Multi-Objective Bio-Inspired Heuristic Cuckoo Search Node Optimization-Based Efficient Routing

The best way of selecting a route is usually to predict traffic between the source and the destination points in VANET routing. The direction of a low traffic density is favored for optimum route choice, as a high traffic density increases the flow of vehicles on the road. For choosing the best path, therefore, an active prediction method is necessary. Route exploration is the central process in the current implementation method.

The proposed algorithm chooses the optimal multipath routing to provide the information efficiently. For the purpose of efficient route detection there is a need to initialize the parameters. To initialize the parameters for the upper and lower band the nest size and mutation probability needs to be calculated. Selecting a proper nest is necessary,

$$\beta = \beta_{max} - (N_{i\,ter}/N_{I\,ter\,(total)} * \beta_{max} - \beta_{min} \qquad (1)$$

Where,

β_{Max} and β_{min} represent the maximum and minimum nest size

$N_{i\,ter}$ represents the current iteration number

$N_{I\,ter\,(total)}$ represents the total iteration number respectively.

As expressed in the equation (1) nest size will decreases with increase in the iteration number. According to MOBHCSNO-ER algorithm mutation probability is associated with the fitness,

$$P_f = \{P_{f\,min} + (P_{f\,max} - P_{f\,min}) \qquad (2)$$

where f=fitness (I)-f $_{min,}$ Which depends on the current quality of Ith solutions; fitness of the population, respectively l; $P_{f\,max}$ and $P_{f\,min}$ represent the maximum and minimum of mutation probability p_a, respectively. From the equation 2 it can be seen that the fitness of the solution gas adjusted, and the mutation probability can generally vary with respect to the amount of the iteration. After the parameter gets initialized the nest position gets analyzed.

$$\alpha_v = (\gamma\,(1+\beta_p))*\sin(\pi*\beta_p/2)/(\gamma(1+\beta_p2)*\beta_p^{\beta_{p-1/2}})*(\beta p-1/2)^{1/(\beta_{p-1/2})} \qquad (3)$$

Where $\sigma_{v,}$ represents the random size of the nest.

The equation (3) can be rewritten as,

$$n_p = rand(\tau_{san}, 1) * (u_b - l_b) + l_b \qquad (4)$$

When n_p represents the random position of the nest. After that the objective function needs to be computed to find out nearest node path. The objective of node localization is to estimate the coordinates of the unknown nodes based on the anchor vehicular nodes. Separably each unknown node and its anchor node can be calculated.

$$\sigma^2 = \gamma^2 * e_{ij}^2 \qquad (5)$$

Where σ^2 is the error variance and e_{ij}^1 is the original distance of the unknown node.

The measured distance between the unknown distance node and the anchor node was represented by using the equation (6)

$$e_{ij'} = e_{ij} = N_{ij} \qquad (6)$$

Where N_{ij} represents the error of the unknown vehicular node. Then the objective function should be calculated which is a combined mean error of the unknown node and the anchor nodes.

$$f(x_i y_i) = 1/n \sum_{j=1}^{n} \frac{(eij - eij')^2}{} \qquad (7)$$

The unknown vehicular node can be estimated by its co-ordinates by running the MOBHCSNO-ER algorithm. When the objective function gets minimized, the unknown short distance path node was easily estimated.

$$Obj_{fn} = -20 * exp\ (-2 * \sqrt{\sum \sigma v})/2exp\ (\sum cos(2\pi * \sigma v)/db)20exp\ (1) \qquad (8)$$

After that the alpha, beta and the gamma positions is updated. After updating it the route can be found. Then the route response is send from the destination, multiple paths will be transmitted to the source. The remaining energy, load and hop count of each node are attached to the reply packet. The origin node then analyses the response packets obtained from several routes and determines the fitness quality for each route on the path.

The highest fitness level and the n path places that chooses the n amount of paths. The routing table of the source node that has according to the fitness values of each of the path in the descending order of the data transmits via the fitness path. Whether the path fails, it is transmitted by the next fitness value path.

Then, if all n paths struggle to transmit information, then it begins to discover and repeat the process according to the algorithm.

If REPLY sends to the neighbor node via destination node until the sends that reaches, the routing table is modified by dragging them up until the recipient vehicular node is hit. The fitness value at sender vehicular node is now determined by applying the mentioned formula in the algorithm 1.

The descending order of the routing table and the data to be stored in the form of the descending order. Now the sender vehicular node actually begins transmitting data according to the path that has the highest fitness value in the routing table of the sender vehicular node, if the path fails the sender sends it with the second-best fitness path, and so on.

Where F is the fitness value of any path received and classified by the source node. Here the residual energy between each of the nodes and the bandwidth distance between each of the pair node can be calculated. The energy, metric delay and shortest route are used to determine fitness. Below is the solution of the suggested methodology. Here the cuckoo's egg is the data packet send by the source vehicular node.

Here the vehicular source along with the data packets is send through the multi objective path that can be delivered to the designator node. Data is passed through the unreliable power or high traffic path is dropped out.

Algorithm 1 (MOBHCSNO-ER)

Input: Vehicle Node Vn, Node coordinates V_c, bound limit V_x, V_y, energy parameter V_e,
Output: Optimized valued φ_p and γ_i (best nest and route path)

119

Step 1: initiates the parameters,
$[V_x, V_y]$ = [Upper Limit lower limit]
Max - iteration max I_{ter}=100;
Lower band $l_b=V_x(1)*(V_c, 2)$;
Upper band u_b=max (V_c, 2);
Best near n_b Dimensions d_b=size (u_b, 2);

Step 2: initialize the nest position n_b,
N= size (n_p)
Beta _ pos β_p=3/2;
$\sigma_v = (\gamma(1=\beta_p))*\sin(\pi*\beta_p/2)/(\gamma(1+\beta_p/2)*\beta_p\beta_p^{-1/2})*(\beta_p=1/2)^{1/(\beta_p-1/2)}$
for j=1: n
s= n_p (j, :)
u= randn (size(s)*σ_v)
v= rand (size(s))
Let compute nest position, n_p=rand (τ_{san}, 1)*(u_b-l_b) + l_b

Modules
Creating the VANET environment
Route discovery
Route request
Route reply
Registration process in the RSU
Vehicular communication using RSU

Creating the VANET Environment

We are going to build the vehicles that are inbuilt with the sensor. Setup the RSU's for the particular coverage area of the vehicles. Build the TA (Trusted Authority) which will check the vehicle entering to particular coverage area and provide authentication to the user.

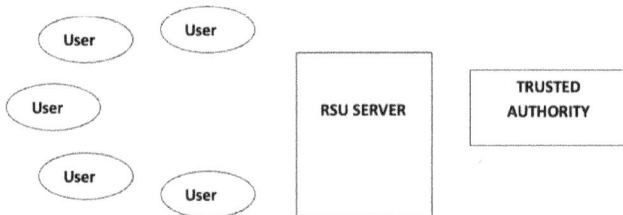

Figure 5. VANET environment

Route Discovery

- If the source vehicle has no route to the destination vehicle, then source vehicle initiates the route discovery in an on-demand fashion
- After generating RREQ, node looks up its own neighbor table to find if it has any closer neighbor vehicle toward the destination vehicle.
- If a closer neighbor vehicle is available, the RREQ packet is forwarded to that vehicle.
- If no closer neighbor vehicle is the RREQ packet is flooded to all neighbor vehicles.
- A destination vehicle replies to a received RREQ packet with a route reply (RREP) packet in only the following three cases:

1) If the RREQ packet is the first to be received from this source vehicle

2) If the RREQ packet contains a higher source sequence number than the RREQ packet previously responded to by the destination vehicle

3) If the RREQ packet contains the same source sequence number as the RREQ packet previously responded to by the destination vehicle, but the new packet indicates that a better quality route is available.

Registration Process in the RSU

- All the users in the VANET should register their details in the RSU.
- After registration the RSU will provide one initial packet key to the user.
- Using this initial packet key, the user will get information about the other nearby vehicles from the TA.

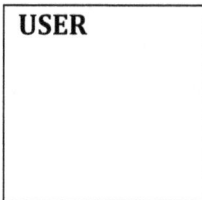

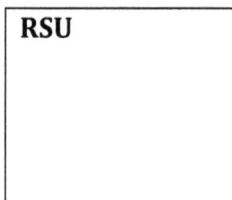

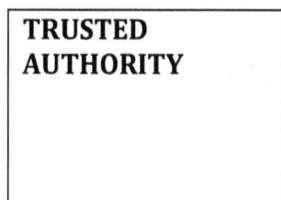

USER	RSU	TRUSTED AUTHORITY

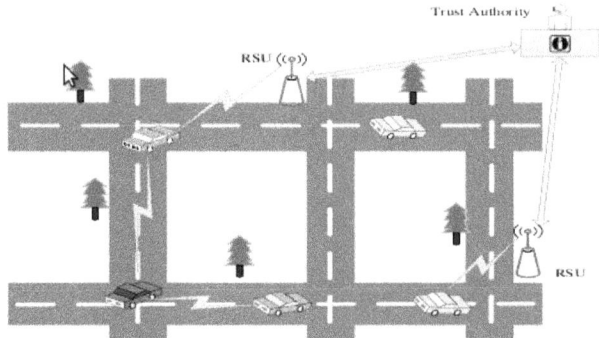

Figure 6. Road map.

Vehicular Communication using RSU

In this module, we are implementing a routing protocol to transfer messages between the vehicles through RSU'S. This communication should be service oriented so that the RSU is exploited from obtaining the various types of data.

RSU is a wireless communication device that uses the DSRC protocol It is a communication medium between vehicle and TA. Vehicles in the same RSU range allowed to exchange secret information.

Vehicle to infrastructure communication is mainly between the vehicle and an RSU fixed on the side of the road. Both vehicle to vehicle and vehicle to infrastructure communications are controlled by short range wireless communication protocol called dedicated short range communication protocol, which improves the overall security and efficiency of the transportation system.

RSU connects to TA over insecure wireless communication media. RSU is not completely trusted, it must be managed and monitored by TA. Vehicle to vehicle communication only through the RSU. Vehicle cannot directly communicate with each other without help of the RSU.

Model Output 1

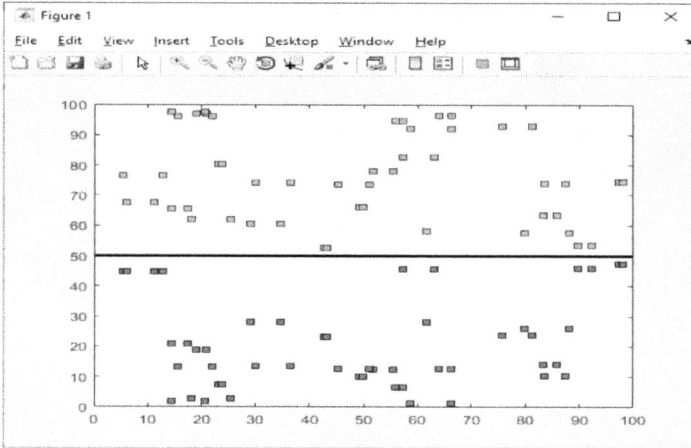

Model Output 2

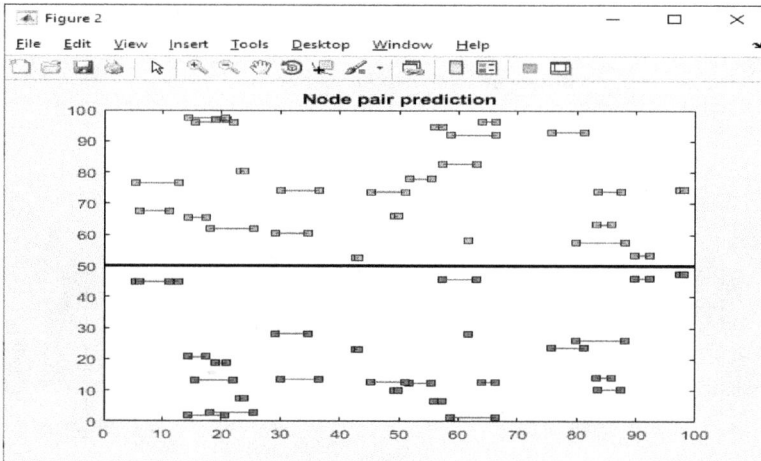

Model Output 3

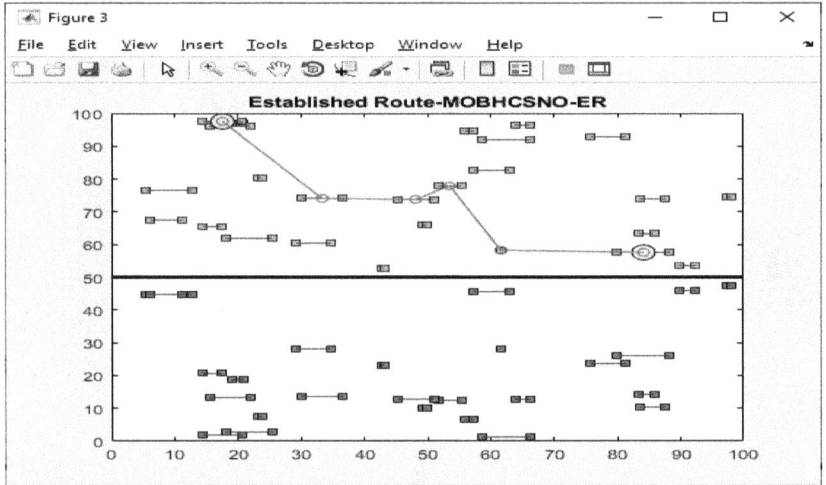

Model Output 4

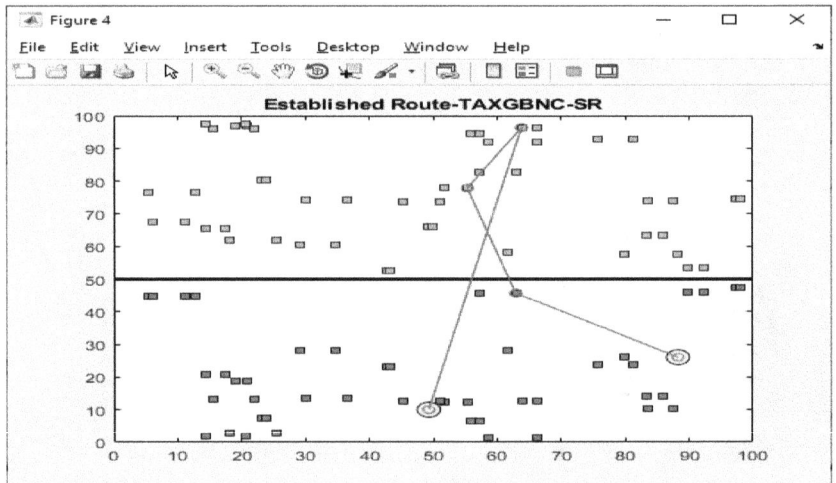

Model Output 5

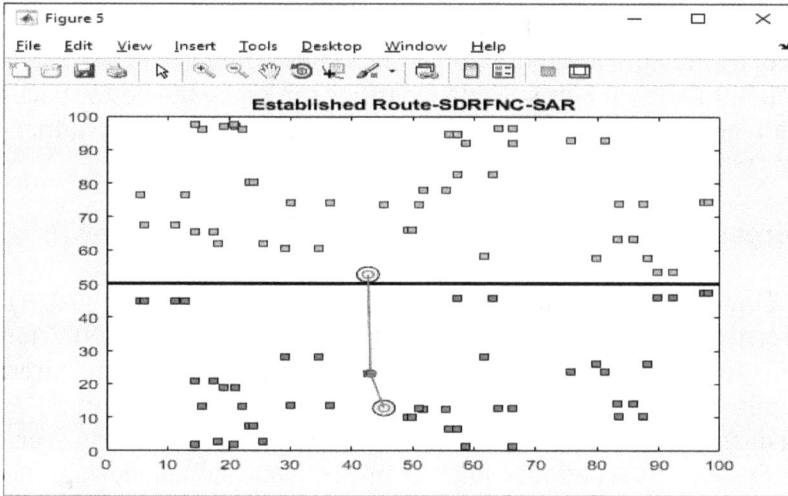

Performance of Proposed Methodology

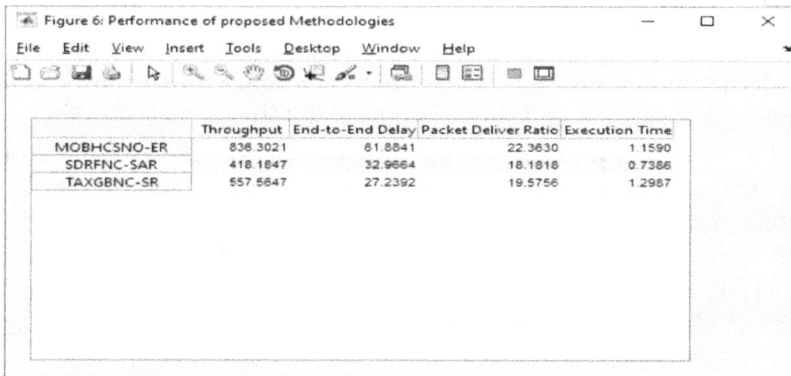

	Throughput	End-to-End Delay	Packet Deliver Ratio	Execution Time
MOBHCSNO-ER	836.3021	61.8841	22.3630	1.1590
SDRFNC-SAR	418.1847	32.9664	18.1818	0.7386
TAXGBNC-SR	557.5647	27.2392	19.5756	1.2987

Conclusion

Routing in VANETs is a complex task in urban environment. The routing issues will occur due to the increase mobility in the nodes. Hence the causes of decreased quality of the services will make the decreased challenging issue. To replace this issue the implementa-

125

tion of Stochastic Discriminant Random Forest Node Classifier and TAXGBNC-SR with the use if the weighted end to end delay approach to be used. Here by using this method a shortest safest path was determined. The simulation results showed that Stochastic Discriminant Random Forest Node Classifier can make the better packet delivery radio, delay, throughput and execution time that outperforms well that will prove the efficiency of the method.

References

[1] Tomar, R.S., Sharma, M.S.P., Jha, S., & Chaurasia, B.K. (2019). Performance Analysis of Hidden Terminal Problem in VANET for safe Transportation system. In harmony Search and Nature Inspired Optimization Algorithm (pp. 1199 -1208). Springer, Singapore.
[2] Loganathan, G. B. (2019). vanet Based Secured Accident Prevention System. International Journal of Mechanical Engineering and Technology, 10(6).
[3] Karimzadeh Motallebiazar, M., Mariano de Souza, A., Zhao, Z., Braun, T., Villas, Z., L., Sargento, S., &Loureire, A.A. (2019). Intelligent Safety Message Dissemination with Vehicle Trajectory Density Prediction in VANETs.
[4] Koti, R.B., & Kakkasageri, M.S. (2019, march). Intelligent Safety Information Dissemination Scheme for V2V Communication in VANETs. In 2019 IEEE International Conference on System, Computation, Automation and Networking (ICSCAN) (pp. 1-6). IEEE.
[5] Jose, a. a., Pramod, A., Philip, G., &George, S.J. (2019). Sybil attack detection in vanet using spidermonkey technique and ecc. International Journal,8(3).
[6] Nguyen, V., Khanh, T.T., Oo, T. Z., Tran, N.H., Huh, E.N., &Hong, C.S.(2019). A Cooperative and Reliable RSU-Assisted IEEE 802.11 P-Based Multi-Channel MAC Protocol for VANETs. IEEE Access, 7, 107576-107590.
[7] Tomar, R.S., Sharma, M.S.P., Jha, S., &Sharma B. (2019). Vehicles Connectivity-Based Communication Systems for Road Transportation Safety. In Soft Computing: Theories and Applications (pp. 483-492). Springer, Singapore.
[8] Li, W., Song,W., Lu, Q., & Yue, C. (2019). Reliable Congestion Control Mechanism for Safety Application in Urban VANETs. Ad Hoc Networks, 102033.
[9] Naja, A., Boulmalf, M., & Essaaidi, M. (2019). Toward a New Broadcasting Protocol to Disseminate Safety Messages in VANET

Environment. In International Conference on Mobile, Secure, and Programmable Networking (pp. 163-172). Springer, Cham.

[10] Benkirane, S. (2019, April). Road Safety against Sybil Attacks Based on RSU Collaboration in VANET Environment. In International Conference on Mobile, Secure, and Programmable Networking (pp. 163-172). Springer, Cham.

[11] Cui, J., Wu, Zhang, J., Xu, Y., &Zhong, H. (2019). An Efficient Authentication Scheme Based on Semi-Trusted Authority in VANETs. IEEE Transaction on Vehicular Technology, 68(3), 2972-2986.

[12] Ramakrishnan, B., Sreedivya, S.R., & Selvi, M. (2015). Adaptive routing protocol based on cuckoo Search algorithm (ARP-CS) for secured vehicular ad hoc network (VANET). International Journal of computer networks and applications (IJCNA), 2(4), 173-178.

[13] Erskine, S. K., &Elleithy, K. M. (2019). Real Time Detection of DoS Attacks in IEEE 802.11 p Using Fog Commuting for a Secure Intelligent vehicular network. Electronics, 8(7), 776.

[14] Limbasiya, T., &Das, D. (2019). ESCBV: energy-efficient and secure communication using batch verification scheme for vehicle users. Wireless Networks, 25(7), 4403-4414.

[15] Hasrouny, H., Samhat, A. E., Bassil, C., & Laouiti, A. (2019, April). A Trusted Group-Based Revocation Process for Intelligent Transportation System. In International Conference on Digital Economy (pp. 133-146). Springer, Cham.

Chapter 11

Enhancing NoC Performance by U-Tolerant Trace Compressor Based Reusing Trace Buffers

S. Ganesh[1], R. Guruprasath[2] and G. Srividhya[2]
[1]Department of Computer Science and Engineering, Study World College of Engineering, Coimbatore, India
[2]Department of Electronics and Communication Engineering, Gnanamani College of Technology, Namakkal, India

Abstract

Design-for-Debug structures such as trace buffers and monitors are usually inserted in Systems-on-Chip to enhance signal visibility. As it is usually inevitable to have unknown 'U' values during silicon debug, trace compressor should be equipped with U-tolerance feature in order not to significantly degrade error detection capability. We propose a scheme AugVC (Augmented Virtual Channel) to reuse trace buffers to augment router buffers, with the objective of improving the overall network performance. Also presents a novel reconfigurable U-tolerant trace compressor, namely U-Tracer, which is able to tolerate as many U-bits as possible in the trace streams while guaranteeing high compression ratio, at the cost of little extra design-for-debug hardware. Output Port directed Virtual Channel (ODVC) that uses a modified virtual channel assignment strategy, on the basis of the designated output port of a network packet. Experimental results on benchmark circuits demonstrate the effectiveness of the proposed technique.

Introduction

Domain Description

Very-large-scale integration (VLSI) is the process of creating an **integrated circuit** (IC) by combining thousands of **transistors** into a single chip. VLSI began in the 1970s when complex **semiconductor** and **communication** technologies were being developed. The **microprocessor** is a VLSI device.

Before the introduction of VLSI technology, most ICs had a limited set of functions they could perform. An **electronic circuit** might consist of a **CPU, ROM, RAM** and other **glue logic**. VLSI lets IC designers add all of these into one chip.

The electronics industry has achieved a phenomenal growth over the last few decades, mainly due to the rapid advances in large scale integration technologies and system design applications. With the advent of very large scale integration (VLSI) designs, the number of applications of integrated circuits (ICs) in high-performance computing, controls, telecommunications, image and video processing, and consumer electronics has been rising at a very fast pace.

VLSI Design Flow

The VLSI IC circuits design flow is shown in the figure below. The various levels of design are numbered, and the blocks show processes in the design flow. Specifications comes first, they describe abstractly, the functionality, interface, and the architecture of the digital IC circuit to be designed.

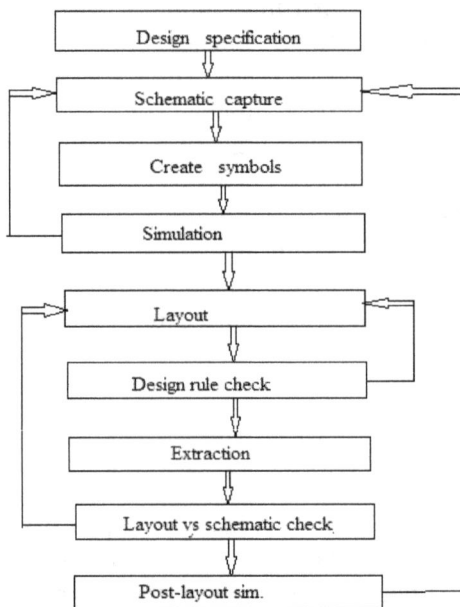

Figure 1. Simplified VLSI design flow.
130

Behavioral description is then created to analyze the design in terms of functionality, performance, compliance to given standards, and other specifications.

RTL description is done using HDLs. This RTL description is simulated to test functionality. From here onwards we need the help of EDA tools. RTL description is then converted to a gate-level netlist using logic synthesis tools. A gate-level netlist is a description of the circuit in terms of gates and connections between them, which are made in such a way that they meet the timing, power and area specifications. Finally, a physical layout is made, which will be verified and then sent to fabrication.

Y Chart

The Gajski-Kuhn Y-chart is a model, which captures the considerations in designing semiconductor devices. The three domains of the Gajski-Kuhn Y-chart are on radial axes. Each of the domains can be divided into levels of abstraction, using concentric rings. At the top level (outer ring), we consider the architecture of the chip; at the lower levels (inner rings), we successively refine the design into finer detailed implementation: Creating a structural description from a behavioral one is achieved through the processes of high-level synthesis or logical synthesis. Creating a physical description from a structural one is achieved through layout synthesis.

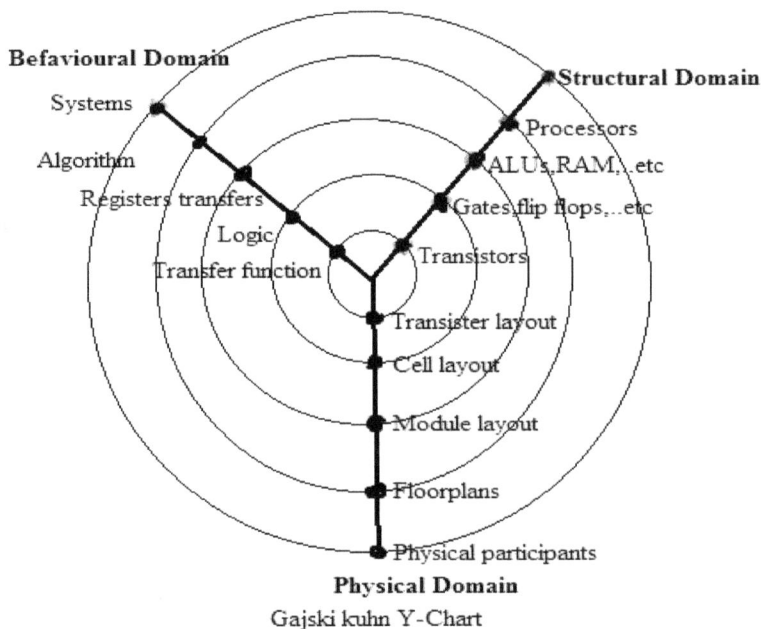

Befavioural Domain

Systems

Algorithm

Registers transfers

Logic

Transfer function

Structural Domain

Processors

ALU,RAM,..etc

Gates,flip flops,..etc

Transistors

Transistor layout

Cell layout

Module layout

Floorplans

Physical participants

Physical Domain

Gajski kuhn Y-Chart

Figure 2. Y chart.

Design Hierarchy-Structural

The design hierarchy involves the principle of "Divide and Conquer." It is nothing but dividing the task into smaller tasks until it reaches to its simplest level. This process is most suitable because the last evolution of design has become so simple that its manufacturing becomes easier. We can design the given task into the design flow process's domain (Behavioral, Structural, and Geometrical). To understand this, let's take an example of designing a 16-bit adder, as shown in Figure 3.

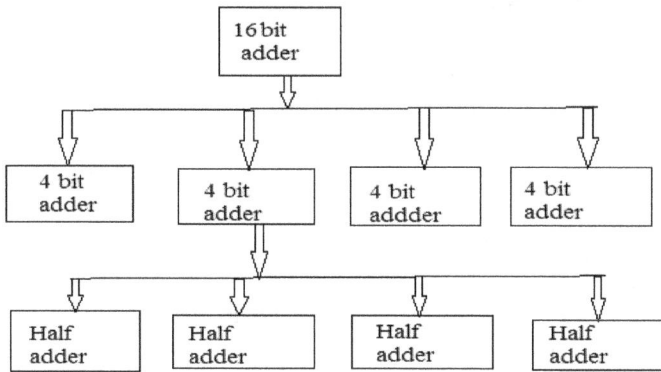

Figure 3. Structural hierarchy of 16 bit adder circuit.

Here, the whole chip of 16 bit adder is divided into four modules of 4-bit adders. Further, dividing the 4-bit adder into 1-bit adder or half adder. 1 bit addition is the simplest designing process, and its internal circuit is also easy to fabricate on the chip. Now, connecting all the last four adders, we can design 4 bit adder and moving on, we can design a 16-bit adder.

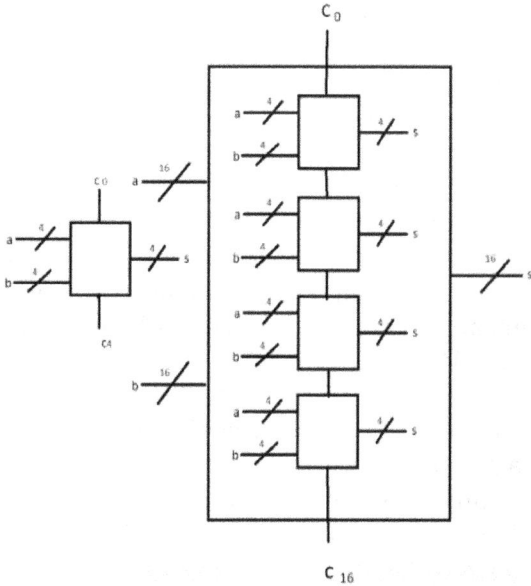

Figure 4. Decomposition of a 4 bit adder.

System–On- Chip (SoC)

System-on-chip (SoC) complexity has increased significantly in terms of both the processor core count as well as the communication among these cores, which makes debugging these systems very challenging. A post-silicon validation phase is usually necessary for ensuring proper system validation. In addition to the traditional focus on component functionality, validation researchers have recently focused on a communication-centric debug methodology to ensure the correctness of on-chip networks. A large part of debug complexity lies in validating the interaction between the system components.

Debugging during post-silicon validation is aided by Design-for-Debug (DFD) hardware which provides visibility into the chip by recording its state. This includes monitors and trace buffers. Monitors are programmed to observe whether the system satisfies various specified conditions and trace buffers are the memory elements that record the router state periodically. This state consists of the flits stored in the router buffers, intended for forwarding to other routers. Router buffers are First-In, First-Out structures that temporarily store the packets that cannot be immediately forwarded due to contention. A flit is a fixed-size unit, one or more of which constitute a packet. There is a tradeoff involved in designing the DFD hardware; increasing this hardware provides higher visibility during debug, but the increased area goes largely unutilized during normal system functioning. A router buffer can be organized either as a single queue as in a wormhole router, or further divided into multiple independent queues called virtual channels (VCs) to avoid head offline blocking as in a virtual channel router. These buffers have a significant impact on the network throughput, especially when the network becomes congested.

Increasing the buffer size reduces the packet drop probability; however, it also increases the on-chip router area and power. Consequently, the buffer design plays an important role in architecting low cost, high performance, and energy efficient on-chip networks.

Chip integration has reached a stage where a complete system can be placed in a single chip. When we say complete system, we mean all the required ingredients that make up a specialized kind of application on a single silicon substrate. This integration has been made possible because of the rapid developments in the field of VLSI designs; this is primarily used in embedded systems. Thus, in simple terms a SoC can be defined as "an IC, designed by stitching together multiple stand-alone VLSI designs to provide full functionality for an application."

While designing an SoC, a vendor may use a library of cores designed by external designers in addition to using cores from in-house libraries. Cores are basically pre-designed models of complex functions termed as Intellectual Property Blocks (IP Blocks), Virtual Components (VC) or simply micros. Since the design of an SoC comprises cores from different sources /vendors, we can say that an SoC is completely heterogeneous, and that is one of the key issues which complicates its design process.

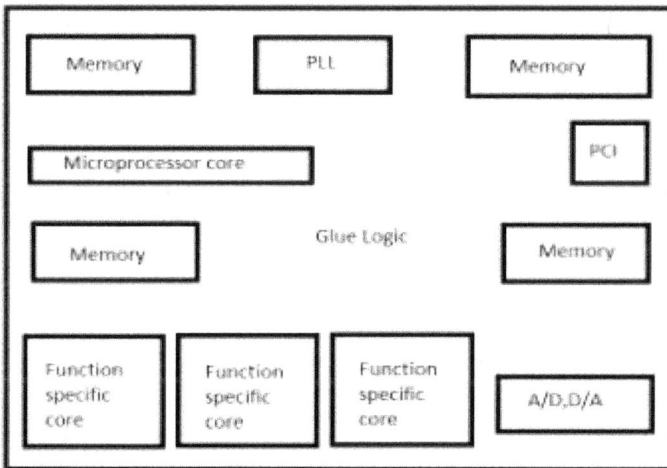

Figure 5. SoC.

A generalized form of today's SoC architectures is depicted in Figure 5. This figure shows the common components used in current SoCs; SRAMS, DRAMS, Flash memory, ROM, DSPs, 2D/3D graphics, and interface cores such as PCI, USB and UART. It should be noted that all these components may belong to different libraries of cores and may belong to different vendors. Also, their organization on the chip depends upon the application they are designed for – Application Specific Integrated Circuit (ASIC). A few examples of today's core

135

based SoCs include GSM mobile phones, single chip digital/videocams, GPS controllers, smart pager ASICs etc.

However, the present SoC architecture doesn't suffice for the future needs, particularly in the terms of their interconnect design due to their poor scalability and inefficiency for handling large number of partners. Hence, from here we move on toward our actual topic of discussion, that is, Network on chips or more commonly called NoCs. A NoC is perceived as a collection of computational, storage and I/O resources on-chip that are connected with each other via a network of routers or switches instead of being connected with point to point wires. These resources communicate with each other using data packets that are routed through the network in the same manner as is done in traditional networks. It is clear from the definition that we need to employ highly sophisticated and researched methodologies from traditional computer networks and implement them on chip. *But why?* In order to elaborate on this question, we have to explore the motivating factors that are compelling the researchers and designers to move toward the adoption of NoC architectures for future SoCs. The area of NoC is still in its infancy, which is one of the reasons why there are various names for the same thing; some call it on-chip networks, some networks on silicon, but the majority agrees upon "Networks on Chips" (NoCs). However, we will be using these terminologies interchangeably throughout our tutorial.

One of the most important trends in computer architecture in the past decade is the move towards multiple CPU cores on a single chip. Common chip multiprocessor (CMP) sizes today range from 2 to 8 cores, and chips with hundreds or thousands of cores are likely to be commonplace in the future. Real chips exist already with 48 cores, 100 cores, and even a research prototype with 1000 cores. While increased core count has allowed processor chips to scale without experiencing complexity and power dissipation problems inherent in larger individual cores, new challenges also exist. One such challenge is to design an efficient and scalable interconnect between cores. Since the interconnect carries all inter-cache and memory traffic (i.e., all data accessed by the programs running on chip), it plays a critical role in system performance and energy efficiency.

Unfortunately, the traditional bus-based, crossbar-based, and other non-distributed designs used in small CMPs do not scale to the medium- and large-scale CMPs in development. As a result, the architecture research community is moving away from traditional centralized interconnect structures, instead using interconnects with distributed scheduling and routing. The resulting Networks on Chip (NoCs) connect cores, caches and memory controllers using packet switching routers, and have been arranged both in regular 2D meshes and a variety of denser topologies. The resulting designs are more network-like than conventional small-scale multicore designs. These NoCs must deal with many problems, such as scalability, routing, congestion, and prioritization that have traditionally been studied by the networking community rather than the architecture community.

While different from traditional processor interconnects, these NoCs also differ from existing large-scale computer networks and even from the traditional multi-chip interconnects used in largescale parallel computing machines. On-chip hardware implementation constraints lead to a different tradeoff space for NoCs compared to most traditional off-chip networks: chip area/space, power consumption, and implementation complexity are first-class considerations. These constraints make it hard to build energy efficient network buffers, make the cost of conventional routing and arbitration a more significant concern, and reduce the ability to over-provision the network for performance. These and other characteristics give NoCs unique properties and have important ramifications on solutions to traditional networking problems in a new context.

Recent work in the architecture community considers bufferless NoCs as a serious alternative to conventional buffered NoC designs due to chip area and power constraints. While bufferless NoCs have been shown to operate efficiently under moderate workloads and limited network sizes (up to 64 cores), we find that with higher-intensity workloads and larger network sizes (e.g., 256 to 4096 cores), the network operates inefficiently and does not scale effectively. As a consequence, application-level system performance can suffer heavily. Through evaluation, we find that congestion limits the efficiency and scalability of bufferless NoCs, even when traffic has locality, e.g., as a result of intelligent compiler, system software, and hardware data mapping techniques.

Unlike traditional large-scale computer networks, NoCs experience congestion in a fundamentally different way due to unique properties of both NoCs and bufferless NoCs. While traditional networks suffer from congestion collapse at high utilization, a NoC's cores have a self-throttling property which avoids this congestion collapse: slower responses to memory requests cause pipeline stalls, and so the cores send requests less quickly in a congested system, hence loading the network less. However, congestion does cause the system to operate at less than its peak throughput, as we will show. In addition, congestion in the network can lead to increasing inefficiency as the network is scaled to more nodes. We will show that addressing congestion yields better performance scalability with size, comparable to a more expensive NoC with buffers that reduce congestion.

Networks on Chips (NoCs)

After realizing the inefficiency of traditional buses in SoCs, in conjunction with so many other factors, the designers of SoCs have come to a cross-road where they meet the computer architecture designers who are always interested in finding dynamic and scalable architectures for building microprocessors. The scalability and wide success of the Internet has attracted the attention of computer architecture as well as SoC designers and influenced them to borrow the idea of using packet based switching networks for the design of future SoC communication infrastructure. It is an understood fact that the actual reason behind success of the Internet and its scalability lies in a well defined protocol stack; the idea was to decouple communication from computation. Packet switched communication not only provides high scalability, but also facilitates reuse of the communication architecture. The two major problems faced by SoC designers – *re-usability and scalability*– can, therefore well be addressed by the adoption of packet switched communication infrastructure for SoC interconnects. Also, from a business point of view, it is important to reduce the design time by adopting reuse not only at the computational level but also reuse of the communication structure. This will in turn lower the time to market new products with ease. Keeping in view this idea of the Internet, many researchers have proposed communication architectures based upon packet switched on chip networks for connecting components in the future SoCs.

Another important aspect of NoCs is that they decouple computation from communication, which is essential for chips that contain billions of transistors. Again, the idea comes from traditional networks such as the Internet, where the communication system is based upon a protocol stack irrespective of the number of the communicating partners. Likewise, the communication infrastructure in NoCs will be designed using a protocol stack which provides well defined interfaces separating communication service usage from service implementation. This means that instead of connecting high level modules (like processors, DSPs, controllers etc.) by routing dedicated wires, they are connected to a network that routes packets between them – as *"Route Packets not wires"*.

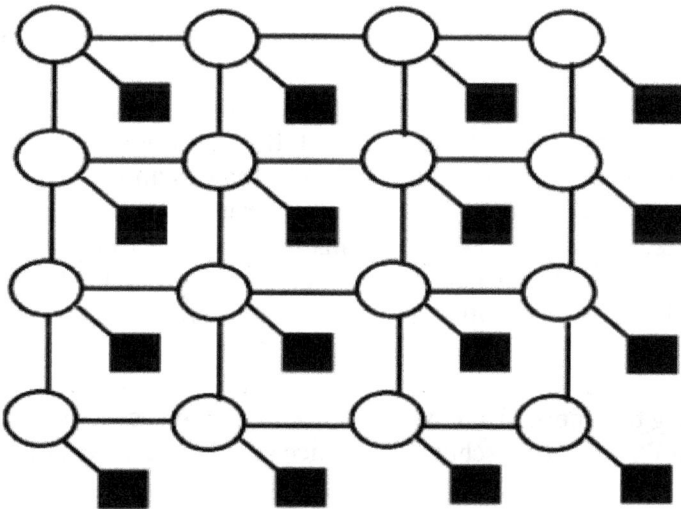

Figuren6. 2D mesh based NoC.

NoC model

Now, since we already know that NoCs is the most appropriate design choice to develop the future SoCs, the next step is to discuss the design of the NoC itself. Since the area of NoCs is really new, it pro-

139

vides us with an opportunity to create things on a clean slate in order to obtain an optimal design. The immense amount of research that is already being conducted in the Internet has been considerably used in defining the structure of NoCs by the researchers. In the following passages we will discuss topologies, protocols, switching and routing mechanisms for NoCs. In the following passages we will discuss proposed topologies, protocols, switching and routing mechanisms for NoCs.

Existing System

In the existing system, the presence of a trace buffer within the router to effectively increase the capacity of the router buffers present at each port is proposed to leverage. The trace buffer is reused to augment the router buffers, with the objective of improving the overall network performance. Since this buffer already exists in the hardware, its reuse incurs minimal area overhead.

In this section, we are to integrate the Trace Buffer, a Design-for-Debug structure already present in the router, into the router architecture, which enables the reclamation and reuse of its storage space for functional purposes. Since all the input ports may not receive packets at the same time, many VCs may go un-utilized even if a demand for VCs exists. We propose to reuse the trace buffer as augmented VC buffers (AugVC). This reuse of the DFD infrastructure reduces the NoC's packet latencies, while also permitting a reduction in the area occupied by the input buffers. In this section we first discuss our proposed router architecture design AugVC.

Following this method, an extension in which we use an output port directed VC allocation scheme to reduce the clock period and router area.

Aug VC Router Architecture

In the AugVC design, this is proposed to reduce the width of the private VCs present at each port and use the trace buffer as a backup storage. To support this new functionality, a new controller is added. The controller first checks the availability of private VCs for an incoming flit and if no space is available in the private VCs, it accommodates those flits in the trace buffer. Similarly, for an outgoing

140

flit, it checks whether a flit of the same packet is present in the trace buffer, in which case, it migrates the flit from the trace buffer to the private VC. Each port can receive a maximum of 1 flit per cycle. Now, a trace buffer of width w flits only serves as a backup storage for w ports because a trace buffer of width w flits allows a maximum of w flits to be written to the trace buffer per cycle.

Let the number of ports be m and the width of the trace buffer be w flits, there are two possible scenarios:
1) m _ w: In this case, the trace buffer can be used as a backup storage for all the m ports, which further allows to reduce the size of all the private VCs.
2) m > w: This scenario is elaborated further. Here, the trace buffer can only be used for w ports and the remaining m⬚ w ports need to use the larger VC sizes to maintain NoC performance.

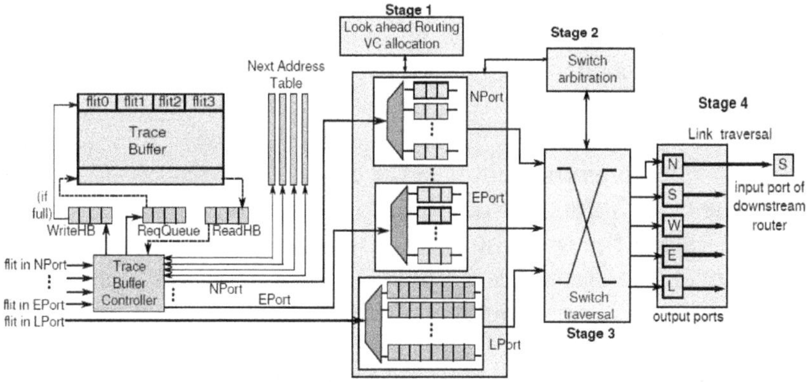

Figure 7. Augmented Virtual Channel (AugVC) router architecture.

In the architecture, an additional component required to support the new functionality are highlighted in blue. For illustration, we consider a router architecture with five input and output ports (North, South, West, East and Local), a flit size of 32 bits, and a trace buffer of width 128 bits. The trace buffer can accommodate only four ports, because a maximum of four flits can be written to the trace buffer per cycle. This allows us to migrate flits from a maximum of four ports. We therefore permit the four nonlocal ports (N, S, W, E) to utilize the trace buffer and consider the local port differently. These can be any four ports. For illustration and evaluation, we have used four non-local ports. The local port has v VCs, each is

141

of size k flits. The other four ports (N, S, E, and W) also have v VCs each, each of size k0 flits, where k0 < k. A relatively small VC size is possible because of the trace buffer being available for backup storage.

Incoming flits from the local core can occupy only one of the k private VC slots available at the local port. Incoming flits that need to be forwarded can occupy the allocated private VC slots (k0) or may be accommodated in the trace buffer if the private VC is full.

We also employ a credit-based flow control system. Each router knows the number of unassigned VCs in each of its downstream routers, the number of available slots in the assigned VCs, as well as the number of available lines in the trace buffer. While allocating a VC for a head flit, the router looks for an unassigned VC in the downstream router, just as in the base router architecture. When forwarding a body or tail flit, the router looks for either a free slot in the corresponding private VC, or an available line in the trace buffer. Whenever a flit leaves a router, the upstream routers are requested to update their credit knowledge accordingly.

When a flit leaves the router, a flit of the same packet is migrated from the trace buffer (if available) to the newly freed slot in the private VC. Since the trace buffer is uniformly shared among all non-local ports, the storage is better utilized than in the case of completely private VCs as employed by the baseline router architecture.

Writing to the Trace Buffer: If there is no space to accommodate an incoming body or tail flit from a non-local port in the private VC, the Trace Buffer Controller stores the flit in a temporary area called the Write Holding Buffer (Write HB). Once the occupancy of the Write HB reaches four, the flits are written to the trace buffer (since writes to the trace buffer are done in blocks of 128 bits or 4 flits). This is done to utilize the trace buffer space efficiently as we may not have flits from all the ports writing into the trace buffer in the same cycle. Thus, a line in the trace buffer may possibly contain data from different VCs of different ports or the same port.

Reading from the Trace Buffer: When a flit exits the private VC, the trace buffer controller checks whether a flit corresponding to that VC is present in the trace buffer, and if so, writes it to the private VC.

To search the trace buffer, the Trace Buffer Controller maintains a linked list of all flits residing in the trace buffer that belongs to the same VC in an auxiliary Next Address Table. The same two values are required to index into the trace buffer and the Next Address Table: line number and position of the entry within the line.

For each VC, a head pointer (<line number, position> tuple) and a tail pointer are maintained, which contain the location of the first and last flits residing in the trace buffer. For a flit f residing at index if in the trace buffer, the entry at index if in the Next Address Table gives the index of the flit following f. The size of Next Address Table is derived from the number of lines and the width (flits per line) of the trace buffer. For example, if the number of lines is n and the trace buffer is w flits wide, then the size of the next address table is $n \_ ((\log 2\ n + \log 2\ w) \_ w)$ bits. The parameters n and w are fixed at design time.

If the head pointer is not null, the trace buffer controller adds a request for the VC in the Request Queue. In one cycle, up to four requests may be added, one from each input port. In every cycle, the head of the Request Queue is processed, and a line is read from the trace buffer and placed in a temporary area called the Read Holding Buffer (Read HB). Further, in every cycle, the Request Queue is scanned to find all those requests that the contents of the Read and the Write HB can satisfy. A flit in the Read HB can be used to satisfy a request only if it is pointed to by the head pointer corresponding to that VC. A flit in the Write HB can be used to satisfy a request only if no flit of that VC is present in the trace buffer. These conditions are necessary for ensuring that the flits are processed in their order of arrival.

A significant fraction of the requests is serviced directly from the Read and Write HBs, reducing the number of trace buffer reads. This is due to the presence of temporal locality; flits that arrive together at the router, tend to enter the trace buffer, and also leave the router together. No changes are made to VC allocation and switch arbitration stages of the pipeline in the AugVC design.

Output Port Dircted VC Assignment (ODVC)

The VC allocation unit and the switch allocation (SA) unit typically form the bottleneck stages of the router pipeline, affecting the clock period of the router. The latencies of these two stages depend on the cardinality of the arbiters used in these stages. During VC allocation, the unit may have to arbitrate as many as v x p requests in a single cycle, where v is the number of VCs per port, and p is the number of ports. This may make the unit reasonably complex, requiring a relatively large clock period. SA is performed in two stages: the first stage selects one of v VCs at each input port, requiring a total of p arbiters of cardinality v : 1. The second stage arbitrates between the winning requests of the first stage to decide who wins each output port. Therefore, a total of p arbiters of cardinality (p – 1) : 1 are required. The two stages of arbitration require a large clock period, and also consume a significant fraction of the router area. Our proposed design reduces the cardinality of the different arbiters, and thus, reduces the overall clock period and the router area. To achieve this, an output directed VC allocation scheme is used.

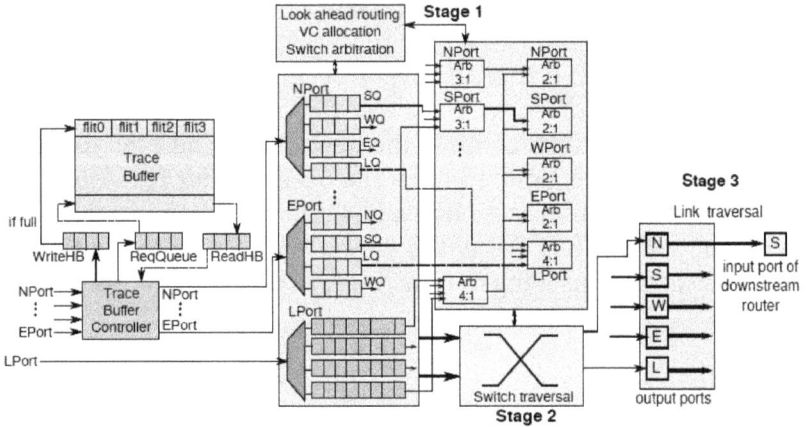

Figure 8. Output Port Directed VC Assignment (ODVC)

Disadvantages
- While the size and organization of the router buffers directly impact network throughput, these buffers also dominate the on-chip router area.

144

- Much larger area overhead is the reuse of the trace buffer for backup storage.
- Reduces the input buffer efficiency.

Proposed System

Reconfigurable Trace Compressor Design

As discussed earlier, while more redundancy is helpful to recover more trace information, the compression ratio is reduced, and it may also involve high control complexity. Consequently, in our trace compressor design, we keep the redundancy ratio to be two. In order to further enhance the capability of the trace compressor, we introduce configurability into our trace compressor design. By doing so, we are able to flexibly change the compressor's structure for each debug run, which enables us to extract more trace information. Figure 9 presents the overall structure of the proposed X-tracer design, which can be configured externally via JTAG interface. Three configuration options are provided to debug engineers. Firstly, the primitive polynomial can be reconfigured for each MISR, and it is implemented by selectively turning on/off the feedback loop from each output. Secondly, the module *Input Order Manipulator* is utilized to change MISR's input order, also individually-controllable.

Finally, we introduce an internal *counter* to determine the cycle number to unload the MISR signatures. Adjustment of the counter value enables us to tradeoff compression ratio and X-tolerant capability. Finally, our proposed trace compressor can be easily reconfigured to simply use a single MISR for trace compression as well, when there is nearly no X-bits and hence redundant tracing is not required.

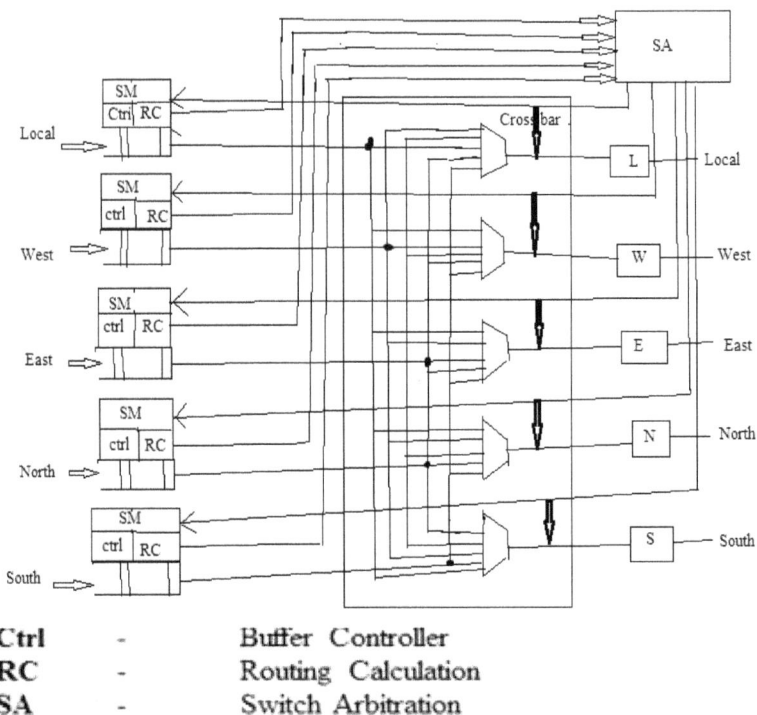

Ctrl	-	Buffer Controller
RC	-	Routing Calculation
SA	-	Switch Arbitration

Figure 9. Reconfigurable Trace Compressor design.

Trace Information Extraction

After acquiring the X-contaminated signatures from the trace buffer in each debug run, we rely on an off-line processing step to extract as many useful trace bits as possible. Given the X-contaminated signatures, we can construct the corresponding X-matrix and employ the X-canceling solution to extract useful trace data. An example X-matrix is shown in the following example, wherein each row corresponds to a MISR observing bit and each column represents a specific X-bit (entry '1' denotes that the corresponding X-bit contaminates the specific observing bit). Next, Gauss-Jordan elimination is utilized to identify all-zero rows, which represent X-canceling combinations. Assuming $I00$, $I02$, $I03$ and $I23$ in Figure 9 are X-bits, one possible solution is shown as follows:

146

$$
\begin{array}{c}
\begin{array}{ccccc} & I_{00} & I_{02} & I_{03} & I_{23} \end{array} \\
\begin{array}{c}
O_7 \\ O_6 \\ O_5 \\ O_4 \\ O_3 \\ O_2 \\ O_1 \\ O_0
\end{array}
\left(
\begin{array}{cccc}
0 & 0 & 1 & 0 \\
1 & 1 & 0 & 0 \\
0 & 0 & 1 & 1 \\
0 & 1 & 0 & 1 \\
1 & 0 & 0 & 0 \\
1 & 0 & 1 & 0 \\
0 & 1 & 0 & 0 \\
0 & 0 & 0 & 1
\end{array}
\right) \rightarrow
\end{array}
\begin{array}{c}
\begin{array}{ccccc} & I_{00} & I_{02} & I_{03} & I_{23} \end{array} \\
\begin{array}{c}
O_3 \\ O_6+O_3 \\ O_7 \\ O_6+O_4+O_3 \\ O_7+O_6+O_5+O_4+O_3 \\ O_7+O_3+O_2 \\ O_6+O_3+O_1 \\ O_6+O_4+O_3+O_0
\end{array}
\left(
\begin{array}{cccc}
1 & 0 & 0 & 0 \\
0 & 1 & 0 & 0 \\
0 & 0 & 1 & 0 \\
0 & 0 & 0 & 1 \\
0 & 0 & 0 & 0 \\
0 & 0 & 0 & 0 \\
0 & 0 & 0 & 0 \\
0 & 0 & 0 & 0
\end{array}
\right)
\end{array}
$$

$$(1)$$

As a test response compaction technique, the objective is to find a solution that X-bits do not contaminate the bits used to fault detection. In silicon debug, however, our objective is to extract as many useful trace bits as possible. Consequently, we cannot directly use to generate our solution.

Since different solutions lead to very different X-canceling combinations, the corresponding extracted trace data may vary significantly. We can in fact employ multiple solutions to extract more trace information for silicon debug. This, however, is a challenging problem because the number of solutions for an X-matrix can be huge (even with the rather constrained Gauss-Jordan elimination procedure). In this section, we propose novel techniques that are able to effectively explore X-canceling combination space and we present an algorithm to maximize the number of extracted trace bits.

X-Canceling Solution Space Exploration

Again, let us take the example shown in Eq. 1 to illustrate our proposed technique for X-canceling solution space exploration. We first select a targeted MISR observing bit (e.g., O6), and move the corresponding row down to the last position. By doing so, we aim at finding the combination of remaining observing bits to cancel the unknown targeted bit. Then, instead of conducting row operations, we perform column operations to reduce the modified Xmatrix to *column echelon form* as follows:

$$
\begin{pmatrix}
O_7 & 0 & 0 & 1 & 0 \\
O_5 & 0 & 0 & 1 & 1 \\
O_4 & 0 & 1 & 0 & 1 \\
O_3 & 1 & 0 & 0 & 0 \\
O_2 & 1 & 0 & 1 & 0 \\
O_1 & 0 & 1 & 0 & 0 \\
O_0 & 0 & 0 & 0 & 1 \\
\hline
O_6 & 1 & 1 & 0 & 0
\end{pmatrix}
\rightarrow
\begin{pmatrix}
1 & 0 & 0 & 0 \\
0 & 1 & 0 & 0 \\
0 & 0 & 1 & 0 \\
0 & 0 & 0 & 1 \\
1 & 0 & 0 & 1 \\
1 & 1 & 1 & 0 \\
1 & 1 & 0 & 0 \\
0 & 1 & 1 & 1
\end{pmatrix}
\tag{2}
$$

With the *column echelon form* of the X-matrix, the first non-zero entry is called a pivot (in bold), and its corresponding row is *pivot row*, which is guaranteed to contain only one non-zero entry. In addition, we define all-zero row as the *free rows*, and the other rows as *stack rows*. According to linear algebra, if the last row (representing the targeted observing bit) is not a pivot row, there must exist at least one combination of the remaining bits to cancel the unknown targeted bit. For the sake of easy explanation, we use a vector S to denote a X-canceling combination, where 1 in S means that the corresponding observing bit is involved in the combination. Based on the above definition for the reduced X-matrix, we define the corresponding bits in the S as *pivot bits, stack bits* and *free bits*. Therefore, an initial solution *Sinit* can be found in the following manner: we first identify non-zero entries on the last row of the reduced X-matrix; then find the pivots on the same column; finally, we fill the related pivot bits in *Sinit* with 1s and left the other as 0s.

For the example in Eq. 2, the initial solution could be *Sinit* = $\{0,1,1,1,0,0,0,1\}$. Next, starting from the initial solution, we try to traverse the solution space and generate new solutions by transforming existing solutions. To guarantee the obtained solution is legal, the transformation obeys three well-defined bit flipping rules: (i). free bits can be freely flipped to generate a new valid solution; (ii). to flip stack bit, we need to further flip all the pivot bits whose corresponding pivots are on the same columns of non-zero entries of stack row correlated with to-be-flipped stack bit. For example, if we flip the fifth bit in *Sinit* , the related stack row is $\{1,0,0,1\}T$ and thus the first and forth pivots bit need be flipped at the same time. In this case, the new solution could be *Ssec* = $\{1,1,1,0,1,0,0,1\}$. (iii). the pivot bits and the last bit in each row cannot be flipped. With the above solution space exploration procedure, the X-canceling combinations can be acquired by simply changing different targeted observing bits.

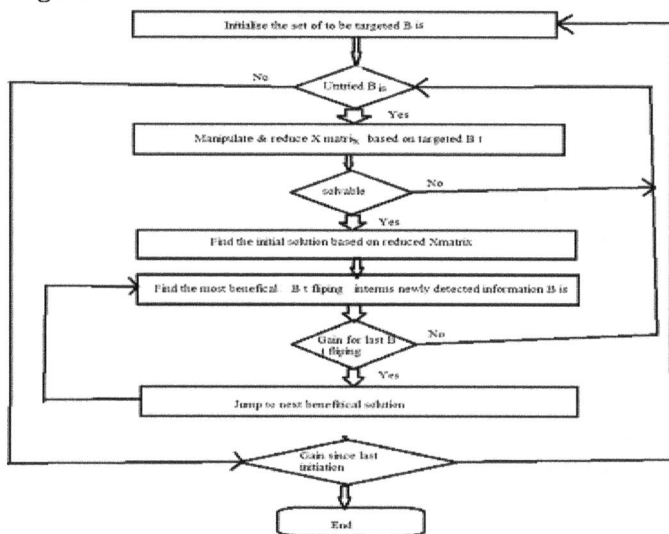

Figure 10. X-Free Information Extraction Algorithm

X-Free Information Extraction

With the flexibly to explore the X-canceling combination solution space, our objective is to extract the maximum number of useful trace data (i.e., those X-free bits that are known). Our proposed algorithm is depicted in Figure 10. It starts from putting all the observing outputs into a to-be-targeted bit set, from which we select one output as the targeted bit a time. Based on the original X-matrix, we first move the targeted bit column to the last position, and then perform row operations to reduce the matrix to *row echelon form*. Next, we try to find the next targeted bit if the reduced matrix is solvable, otherwise we try to restore the initial solution. Then, we search the solution space by greedily flipping the most beneficial bit, in which the *gain* is defined as the increased number of extracted information bits, and the search procedure is stopped when there is no more *gain*. After all the targeted bits have been tried, we check whether these is still *gain* since the last initialization. The whole procedure is re-initialized if it is not zero, otherwise it is terminated.

149

Advantages

- Reuse improves network performance with reduced area.
- Reduces the cardinality of the d ifferent arbiters, and reduces the clock period and area of the router.
- The buffer architecture plays an important role in designing low cost, high performance, and energy efficient on-chip networks.
- U-bits as possible in the trace, guaranteeing high compression ratio, at the cost of little extra design-for-debug hardware.

Applications

- Manufacturing cell phones, MP3 players, computer printers and peripherals.
- IC components used in transportation safety and comfort
- Network device management systems

FPGA Implementation

Spartan 3

The Spartan-3 family of Field-Programmable Gate Arrays is specifically designed to meet the needs of high volume, cost-sensitive consumer electronic applications. The eight-member family offers densities ranging from 50,000 to five million system gates. The Spartan-3 family builds on the success of the earlier Spartan-IIE family by increasing the amount of logic resources, the capacity of internal RAM, the total number of I/Os, and the overall level of performance as well as by improving clock management functions.

These Spartan-3 FPGA enhancements, combined with advanced process technology, deliver more functionality and bandwidth per dollar than was previously possible, setting new standards in the programmable logic industry. Because of their exceptionally low cost, Spartan-3 FPGAs are ideally suited to a wide range of consumer electronics applications including broadband access, home networking, display/ projection and digital television equipment. The Spartan-3 family is a superior alternative to mask programmed ASICs. FPGAs avoid the high initial cost, the lengthy development cycles, and the inherent inflexibility of conventional ASICs. Also, FPGA programmability permits design upgrades in the field with no hardware replacement necessary, an impossibility with ASICs.

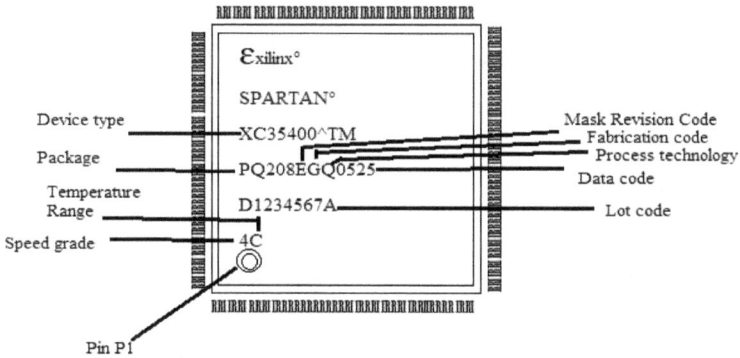

Figure 11. Spartan-3 QFP Package XC3S400-4PQ208C.

Architectural Overview

The Spartan-3 family architecture consists of five fundamental programmable functional elements:
1. Configurable Logic Blocks (CLBs) contain RAM-based Look-Up Tables (LUTs) to implement logic and storage elements that can be used as flip-flops or latches. CLBs can be programmed to perform a wide variety of logical functions as well as to store data.
2. Input/ Output Blocks (IOBs) control the flow of data between the I/O pins and the internal logic of the device. Each IOB supports bidirectional data flow plus 3-state operation. Twenty-six different signal standards, including eight high-performance differential standards are available. Double Data-Rate (DDR) registers are included. The Digitally Controlled Impedance (DCI) feature provides automatic on-chip terminations, simplifying board designs.
3. Block RAM provides data storage in the form of 18-K bit dual-port blocks.
4. Multiplier blocks accept two 18-bit binary numbers as inputs and calculate the product.

5. Digital Clock Manager (DCM) blocks provide self-calibrating, fully digital solutions for   distributing, delaying, multiplying, dividing and phase shifting clock signals.

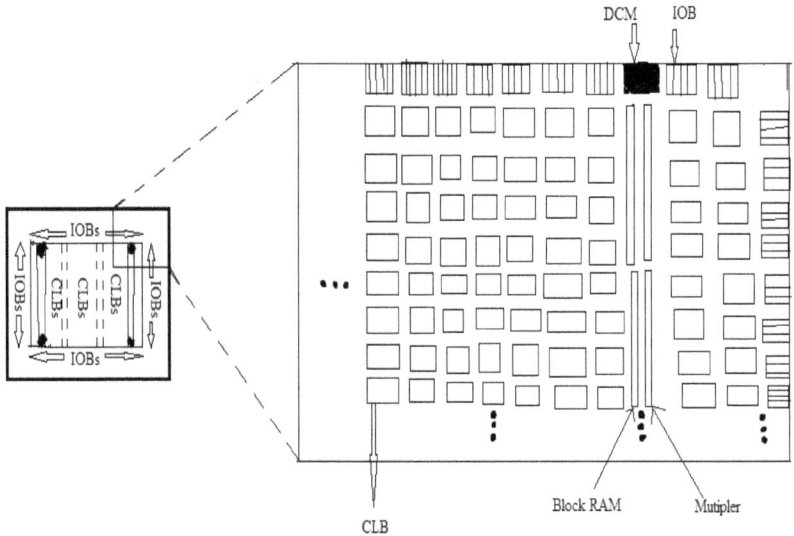

Figure 12. Spartan-3 family architecture.

A ring of IOBs surrounds a regular array of CLBs. The XC3S50 has a single column of block RAM embedded in the array. Those devices ranging from the XC3S200 to the XC3S2000have two columns of block RAM.

The XC3S4000 and XC3S5000 devices have four RAM columns. Each column is made up of several 18-Kbit RAM blocks; each block is associated with a dedicated multiplier. The DCMs are positioned at the ends of the outer block RAM columns. The Spartan-3 family features a rich network of traces and switches that interconnect all five functional elements, transmitting signals among them. Each functional element has an associated switch matrix that permits multiple connections to the routing.

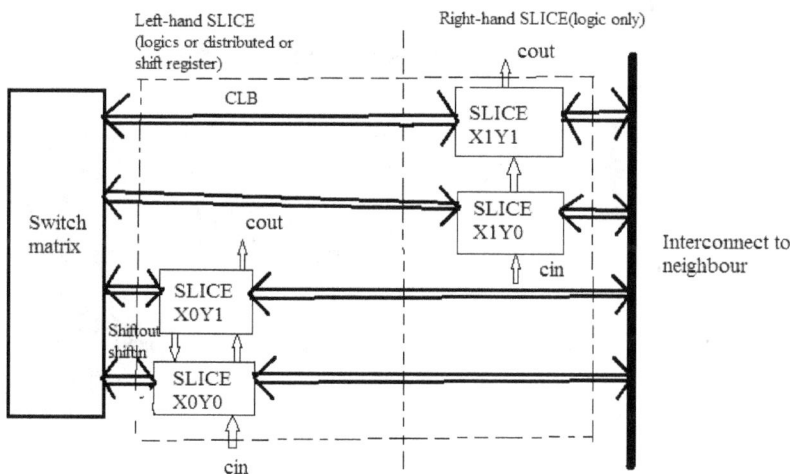

Figure 13. Arrangement of slices within the CLB

The Configurable Logic Blocks (CLBs) constitute the main logic re-source for implementing synchronous as well as combinatorial cir-cuits. Each CLB comprises four interconnected slices as shown in Figure 13. These slices are grouped in pairs. Each pair is organized as a column with an independent carry chain.

The nomenclature that the FPGA Editor — part of the Xilinx devel-opment software — uses to designate slices is as follows: The letter 'X' followed by a number identifies columns of slices. The 'X' num-ber counts up in sequence from the left side of the die to the right. The letter 'Y' followed by a number identifies the position of each slice in a pair as well as indicating the CLB row. The 'Y' number counts slices starting from the bottom of the die according to the sequence: 0, 1, 0, 1 (the first CLB row); 2, 3, 2, 3 (the second CLB row); etc. Fig. 5.3 shows the CLB located in the lower left-hand cor-ner of the die. Slices X0Y0 and X0Y1 make up the column-pair on the left whereas slices X1Y0 and X1Y1 make up the column-pair on the right. For each CLB, the term "left-hand" (or SLICEM) indicates the pair of slices labeled with an even 'X' number, such as X0, and the term "right-hand" (or SLICEL) designates the pair of slices with an odd 'X' number, e.g., X1.

Simulation Results

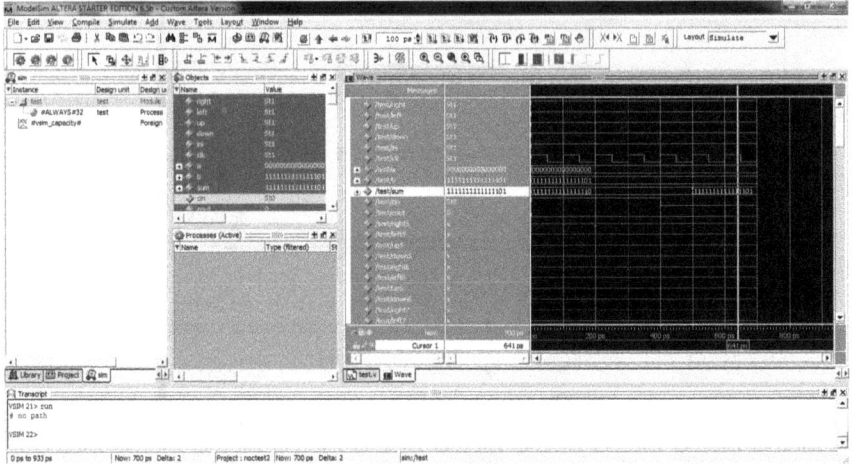

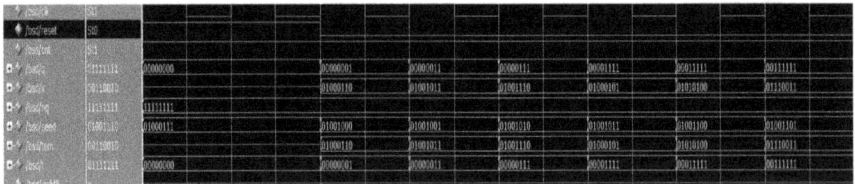

Conclusion

As network-on-chip performance is directly related to router buffer configuration, the buffer architecture plays an important role in designing low cost, high performance, and energy efficient on-chip networks. In this work, Proposed a reconfigurable trace data compressor design with explicit redundancy for silicon debug, in the presence of many X-bits during signal tracing in re used buffers. The proposed *X-tracer* design, together with the novel algorithms to extract useful trace data out of contaminated trace signatures, facilitates to obtain as much trace information as possible while guaranteeing high compression ratio, as demonstrated in our experimental results.

References

[1] Neetu Jindal, Shubhani Gupta, Divya PraneethaRavipati, "Enhancing Network-on-Chip Performance by Reusing Trace Buffers" DOI10.1109/TCAD.2019.2907909, IEEE Transactions on Computer-Aided Design of Integrated Circuits and Systems.

[2] P. Taatizadeh and N. Nicolici, "Emulation infrastructure for the evaluation of hardware assertions for post-silicon validation," IEEE Transactions on Very Large Scale Integration (VLSI) Systems, vol. 25, no. 6,pp. 1866–1880,2017.

[3] C. Li, D. Dong, X. Liao, J. Wu, and F. Lei, "Rob-router: Low latency network-on-chip router microarchitecture using reorder buffer," in High-Performance Interconnects (HOTI), 2016 IEEE 24th Annual Symposium. IEEE, 2016, pp. 68–78.

[4] A. Basak, S. Bhunia, and S. Ray, "Exploiting design-for-debug for Flexible SoC security architecture," in Proceedings of the 53rd Annual Design Automation Conference (DAC). ACM, 2016, p. 167.

[5] S. A. R. Jafri, H. B. Sohail, M. Thottethodi, and T. Vijaykumar, "apslip: A high-performance adaptive-effort pipelined switch allocator," 2013.

[6] Y. Xu, B. Zhao, Y. Zhang, and J. Yang, "Simple virtual channel allocation for high throughput and high frequency on-chip routers," in High Performance Computer Architecture (HPCA), 2010 IEEE 16th International Symposium on. IEEE, 2010, pp. 1–11.

[7] Y.-H. Kao, N. Alfaraj, M. Yang, and H. J. Chao, "Design of high-radix clos network-on-chip," in Proceedings   of the 2010 Fourth ACM/IEEE International Symposium on Networks-on-Chip. IEEE Computer Society, 2010, pp. 181–188.

[8] K. Goossens   B. Vermeulen, and A. B. Nejad, "A high-level debug environment for communication-centric debug," in Proceedings of the Conference on Design, Automation and Test in Europe (DATE). European Design and Automation Association, 2009, pp. 202–207.

[9] B. Vermeulen and K. Goossens, "A network-on-chip monitoring infrastructure for communication-centric debug of embedded multi-Processor socs," in VLSI Design, Automation and Test,2009. VLSI-DAT'09. International Symposium on. IEEE, 2009, pp. 183–186.

[10] A. M. Gharehbaghi and M. Fujita, "Transaction-based debugging of system-on-chips with patterns," in Computer Design, 2009. ICCD 2009. IEEE International Conference on. IEEE, 2009, pp. 186–192.

[11] Y. Hoskote, S. Vangal, A. Singh, N. Borkar, and S. Borkar, "A 5-GHz mesh interconnect for teraflops processor," IEEE Micro, vol. 27, no. 5, pp. 51–61, 2007.

[12] C. Ciordas, K. Goossens, T. Basten, A. Radulescu, and A. Boon, "Transaction monitoring in networks on chip: The on-chip run-time perspective," in Industrial Embedded Systems, 2006. IES'06. International Symposium on. IEEE, 2006, pp. 1–10.

[13] J. Hu and R. Marculescu, "Application-specific buffer space allocation for networks-on-chip router design," in Proceedings of the 2004 IEEE/ACM International conference on Computer-aided design (ICCAD). IEEE Computer Society, 2004, pp. 354–361.

[14] C. Ciordas, T. Basten, A. Radulescu, K. Goossens, and J. Meerbergen, "An event-based network-on-chip monitoring service," in High-Level Design Validation and Test Workshop (HLDVT), 2004. Ninth IEEE International. IEEE, 2004, pp. 149–154.

[15] H. Wang, L.-S. Peh, and S. Malik, "Power-driven design of router microarchitectures in on-chip networks," in Proceedings of the 36th annual IEEE/ACM International Symposium on Microarchitecture. IEEE Computer Society, 2003, p. 105.

[16] T. T. Ye, L. Benini, and G. De Micheli, "Analysis of power consumption on switch fabrics in network routers," in Design Automation Conference, 2002. Proceedings. 39th (DAC). IEEE, 2002, pp. 524–529.

[17] S. Kumar, A. Jantsch, J.-P. Soininen, M. Forsell, M. Millberg, J. Oberg, K. Tiensyrja, and A. Hemani, "A network on chip architecture and design methodology," in VLSI, 2002 (ISVLSI). Proceedings. IEEE Computer Society Annual Symposium on. IEEE, 2002, pp. 117–124.

[18] M. Galles, "Spider: A high-speed network interconnect," IEEE Micro, vol. 17, no. 1, pp. 34–39, 1997.

[19] W. J. Dally, "Virtual-channel flow control," IEEE Transactions on Parallel and Distributed systems (TPDS), vol. 3, no. 2, pp. 194–205, 1992.

[20] M. J. Karol, K. Y. Eng, and H. Obara, "Improving the performance of input-queued atm packet switches," in INFOCOM'92. Eleventh Annual Joint Conference of the IEEE Computer and Communications Societies, IEEE. IEEE, 1992, pp. 110–115.

Chapter 12

An Advanced Encryption Scheme for Securing Colour Image Using Elliptical Curve Cryptography with Hillcipher

S. Ganesh[1] and S. Kannadhasan[2]
[1]Department of Computer Science and Engineering, Study World College of Engineering, Coimbatore, India
[2]Department of Electronics and Communication Engineering, Study World College of Engineering, Coimbatore, India

Abstract

Image encryption is rapidly increased recently by the increasing use of the internet and communication media. Sharing important images over unsecured channels is liable for attacking and stealing. Encryption techniques are the suitable methods to protect images from attacks. Hill cipher algorithm is one of the symmetric techniques, it has a simple structure and fast computations, but weak security because sender and receiver need to use and share the same private key within a non-secure channel. A new image encryption technique that combines Elliptic Curve Cryptosystem with Hill Cipher (ECC-HC) has been proposed in this system to convert Hill Cipher from symmetric technique to asymmetric one and increase its security and efficiency and resist the hackers. Self-invertible key matrix is used to generate encryption and decryption secret key. So, no need to find the inverse key matrix in the decryption process. A secret key matrix with dimensions 4 × 4 will be used as an example in this work.

Introduction

Cryptography is one of the main scientific methods that are utilized to shield images from foes and increment the security of interchanges. Encryption is finished by the sender to change over the first grayscale image to scrambled image before sending it by means of the web to the next client (receiver). Unscrambling is finished by the collector to restore the figured image back to the first image. Symmetric (private key) and lopsided (public key) encryption procedures are two gatherings of cryptography. In symmetric encryption, a similar key (private key) is utilized for both encryption and decoding forms, while in lopsided encryption the sender utilizes a private key unique in relation to the collector's private key and each gathering produces general society and mystery key independently in the wake of concurring on the elliptic Curve space parameters. Both sender and recipient are trading their open keys. Elliptic Curve cryptography (ECC) is one of the compelling open key cryptography techniques. ECC takes a shot at a little key size with a little measure of memory and low force contrasted with different frameworks like RSA, Hill Cipher calculation is one of the symmetric methods; it has high throughput, rapid, and basic structure.

Another encryption system has been proposed right now consolidate Elliptic Curve Cryptosystem (ECC) with Hill Cipher (HC) strategy to fortify the security and produce another methodology (ECCHC). The new methodology utilizes ECC to create the private and open keys, and afterward both sender and collector can deliver the mystery key with no compelling reason to share it through the web or unbound correspondence channel. One of the primary disadvantages in Hill Cipher calculation is that the reverse of the key framework doesn't generally exist. In this way, if the key lattice isn't invertible, the decoding procedure is impossible, and the collector can't get the first information. This paper keeps away from this issue by utilizing the self-invertible key framework (the key lattice is self-invertible if [K = K-1]) which lessens the computational procedure needs during the decoding procedure to process key grid converse. Both sender and recipient develop the self-invertible key lattice and use it for encryption and decoding with no compelling reason to produce the backwards of the key network

Authentication

The process of identifying an individual usually based on key. In security systems, authentication is distinct from authorization, which is the process of giving the key.

Conventional Encryption

Conventional Encryption is referred to as symmetric encryption or single key encryption. It was the only type of encryption in use prior to the development of public-key encryption. Conventional encryption can further be divided into the categories of classical and modern techniques. The hallmark of the classical technique is that the cipher or the key to the algorithm is shared i.e. known by the parties involved in the secured communication. So there are two types of cryptography: secret key and public key cryptography. In secret key same key is used for both encryption and decryption. In public key cryptography each user has a public key and a private key. In this chapter a very view of conventional cryptography is presented. In the section 2.8 a new method of cipher developed by author has been presented.

What is Cryptography?

Cryptography is the study of Secret (crypto-)-Writing (-graphy). It is the science or art of encompassing the principles and methods of transforming an intelligible message into one that is intelligible and then transforming the message back to its original form. As the field of cryptography has advanced; cryptography today is assumed as the study of techniques and applications of securing the integrity and authenticity of transfer of information under difficult circumstances.

Today's cryptography is more than encryption and decryption. Authentication is as fundamentally a part of our lives as privacy. We use authentication throughout our everyday lives when we sign our name to some document and for instance and, as we move to world where our decisions and agreements are communicated electronically, we need to have electronic techniques for providing authentication. Cryptography provides mechanisms for such procedures.

A digital signature binds a document to the possessor of a particular key, while a digital timestamp binds a document to its creation of a particular time. These cryptographic mechanisms can be used to control access to shared disk drive, a high security installation, or a pay-per-view TV channel. The field of cryptography encompasses other uses as well. With just a few basic cryptographic tools, it is possible to build elaborate schemes and protocols that allow us to pay using electronic money, to prove we know certain information without revealing the information itself, and to share quantity in such a way that a subset of the shares can reconstruct the set. While modern cryptography is growing increasingly diverse, cryptography is fundamentally based on problems that are difficult to solve. The problem may be difficult because its solution requires knowledge such as decrypting an encrypted message or signing some digital document. Cryptographic systems are generally classified along three independent dimensions:

1. Type of operations used for transforming plaintext to cipher text. All encryption algorithms are based on two general principles. Those are substitution, in which each element in the plain text is mapped into another element and transposition in which elements in the plaintext are rearranged. The fundamental requirement is that no information be lost. Most systems referred to as product systems, involved multiple stages of substitution and transposition.
2. The number of keys used: If sender and receiver use the same key, the system is referred to as symmetric, single key or secret key conventional encryption. If the sender and the receiver each uses a different key the system is referred to as asymmetric, two key, or public-key encryption.
3. The way in which the plaintext is processed: A block cipher processes the input on block of elements at a time, producing an output block for each input block. A stream cipher processes the input elements continuously, producing output one element at a time, as it goes along.

Steganography, hiding one message inside another, is an old technique that is still in use. For example, a message can be hidden inside a graphics image file by using the low order bit of each pixel to encode the message. The visual effect of these tiny changes is probably too small to be noticed by the user. The message can be hidden further by compressing it or by encrypting it with a conventional cryptosystem. Unlike conventional cryptosystems, where we assume the attacker knows everything about the cryptosystem except for the secret key,

Steganography relies on the secrecy of the method of hiding for its security. If eve does not recognize the message as cipher text, then she is not likely to attempt to decrypt it.

Conventional Encryption Model

A conventional encryption model can be illustrated as assigning Xp to represent the plaintext message to be transmitted by the originator. The parties involved select an encryption algorithm represented by E. The parties agree upon the secret key represented by K. The secret is distributed in a secure manner represented by SC. Conventional encryption's effectiveness rests on keeping the key secret. Keeping the key secret rests in a large on key distribution methods. When E processes Xp and K, Xc is derived. Xc represents the cipher text output, which will be decrypted by the recipient. Upon receipt of Xc, the recipient uses a decryption algorithm represented by D to process Xc and K back to Xp. This is represented in the figure. In conventional encryption, secrecy of the encryption and decryption algorithm is not needed.

Key Distribution-Conventional Cryptography

The strength of any cryptographic system rests on the key distribution technique. Conceivably for two parties: A could select a key and hand delivers it to B, or A and could rely on a trusted courier. If A and B have an established secure connection, they could exchange a key via encrypted messaging, or if A and B have an established secure connection via a third party C, C could provide this trusted courier service. The first two situations could provide a logistical nightmare if the number of communicating parties increase, the number of discrete keys needed also increases exponentially. In option, weaknesses are revealed, that if an attacker ever is able to compromise any one key it can compromise all the keys i.e. attacker can masquerade as the third party. The fourth option demonstrates a Key Distribution Center (KDC) which is largely adopted. A KDC is responsible for securely delivering unique key pair to its clients. It is also responsible for key management. A key management center uses a hierarchy of keys to provide authentication, integrity, no repudiation, and confidentiality to its users. The hierarchy of keys consists of session keys, which are used for logical connection between end users. The session keys are encrypted by a master key which is shared by the KDC and an end user. Consider this classic key distribution scenario: A uses his secret key to request a session key from KDC to establish a logical connection with B. The request includes both the identities of A and B and a unique identifier for the transaction. A is a contrivance invented or used for this, singular occasion. The replies to A with encrypted message which contains the requested one-time session key and the original message with the nonce. The original message is used to verify the reply's integrity. Then once is used by the requestor to verify that the returned message is not a replay of an older request.

The reply also contains two items relating to B. They are the one-time session key, to be used for by session and an identifier for A. Both these items are encrypted with the master key shared by KDC and B. The items are used to authenticate A and B. Party A forwards to B the information that originated at KDC, that was encrypted with B's master key. This gives the process its integrity.

For B to be authenticated to A the following steps should occur: Party B now uses the logical connection created by the shared session key to send a half defined nonce to A. Upon receipt of B defined nonce, A performs a mathematical function on the nonce; A performs a mathematical function on the nonce and returns result to B through the logical connection. The keys have to manage across KDC domains. Keys issued to an entity by one KDC have to validate by the issuer before they can be accepted by an entity serviced by a different issuer. The issuers have to collaborate on acceptable method of authenticating inter-domain transaction. Currently, KDCs use a hierarchy for key sharing. Each local KDC negotiates keys for its subscribers through a global KDC. Session generated must have a finite lifetime. Keys are exchanged frequently to prevent an opponent from having a large amount of data encoded with the same key. This brings us to the subject of cryptanalysis.

Cryptanalysis

Code making involves the creation of encryption products that provide protection of confidentiality. Code breaking involves defeating this protection by some means other than the standard decryption process used by an intended recipient. Five scenarios for which code breaking is used. They are: ensuribg accessibility, spying on opponents, selling cracking products and services, pursuing the intellectual aspects of code breaking and testing whether one's codes are strong enough. Cryptanalysis is the process of attempting to discover either the plaintext X_p or the key K. Discovery of the encryption is the most desired one as with its discovery all the subsequent messages can be deciphered. Therefore, the length of encryption key, and the volume of the computational work necessary pro-

162

vides for its strength i.e. resistance to breakage. The longer the key, the stronger the protection, the more brute force is needed. Neither conventional encryption nor public key encryption is more resistant to cryptanalysis than the other. All that the user of an encryption algorithm can strive for is an algorithm that meets one or both of the following criteria: the cost of breaking the cipher exceeds the value of the encrypted information, the time required to break to exceeds the normal lifetime of the information.

Classical Encryption Techniques

The technique enables us to illustrate the basic approaches to conventional encryption today. The two basic components of classical ciphers are substitution and transposition. Then other systems described that combines both substitution and transposition.

Substitution Techniques

In this technique letters of plaintext are replaced by or by numbers and symbols. If plaintext is viewed as a sequence of bits, then substitution involves replacing plaintext bit patterns with cipher text bit patterns.

Caesar Cipher

Caesar Cipher replaces each letter of the message by a fixed letter a fixed distance away e.g. uses the third letter on and repeatedly used by Julius Caesar.

For example:

Plaintext: I CAME I SAW I CONQUERED
Cipher text: L FDPH L VDZ L FRQTXHUHG
Mapping is:
ABCDEFGHIJKLMNOPQRSTUVWXYZ
DEFGHIJKLMNOPQRSTUVWXYZABC

We can describe the Cipher as:

Encryption: C=E (P)=(P+3)mod 26-------

Decryption: P=D(C) =(C-3)mod26--------

Mono-alphabetic Cipher

The Caesar cipher uses only 26 rotations out of the 26 permutations on the alphabet. The mono-alphabetic cipher uses them all. A key k is an arbitrary permutation of the alphabet. Ek(m) replaces each letter of a of m by k(a) to yield c. To decrypt Dk(c) replaces each letter b of c by k-inverse (b). The size of the key space is 26!> 2 pow(74), which is too large for a successful brute force attack. However, mono-alphabetic ciphers can be easily broken-down using letter frequency analysis, given a long enough message. Because each occurrence of a letter a in the message is replaced by the same letter k(a), the most frequently occurring letter of m will correspond with the most frequently occurring letter of c. while jack might know what the most frequently occurring letter of m is, if the message is long enough and she knows that it is English, then it is quite likely that the most frequently occurring letter in m is one of the most frequently occurring letters in English i.e. 'e' or maybe 't'. She can assume then that the most frequent letter 'b1' in c is 'e', the next most frequent letter b2 in 't'., and so forth. Of course, not all these guesses will be correct, but the number of likely candidates for each cipher text letter is greatly reduced. Moreover, many wrong guesses can be quickly discarded even without constructing the entire trial key because they lead to unlikely letter combinations.

Playfair Cipher

The Playfair is a substitution cipher bearing the name of the man who popularized but not created it. The method was invented by Sir Charles Wheatstone, in around 1854; however, he named it after his friend Baron Playfair. The Playfair Cipher was developed for telegraph secrecy, and it was the first literal digraph substitution cipher. This method is quite easy to understand and learn but not easy to break, because you would need to know the "keyword" to decipher   the code. The system functions on how letters are positioned in a 5*5 alphabet matrix. A "KEYWORD" sets the pattern of letters with the other letters the cells of the matrix in alphabetical order (I and j are usually combined in one cell). For instance, suppose we use a keyword Charles then matrix would look like this:

164

c h a r l
e s b df

g i/j k m no p q tu
v w x y z

Now supposing the message to be enciphered here is "the scheme really works". First of all, the plaintext is divided into two letter groups. If there are double letters occurring, in the message, either an 'x' will be used to separate the double letters or an'x' will be added to make a two-letter group combination. In our example, the phrase becomes: Enciphered text: th es c hem er ea lx ly wo rk sx.

Each of the above two letter combinations will have 3 possible relationships with each other in the matrix: they can be in the same column, same row, or neither. The following rules for replacement should be used:

If two letters are in the same column of the matrix, use letter below it as the cipher text (columns are cyclical).
If two letters are in the same row of the matrix, use letter to the right as the cipher text (columns are cyclical).
If neither the same column or row, then each are exchanged with the letter at the intersection of its own row and the other column. From our example:

Plaintext: thes ch em er lx ly wo rk sx
Ciphertext: pr sb ha dg bc az rz vp ambw
For deciphering, the rules are exact opposite.

Hill Cipher

The core of Hill-cipher is matrix manipulations. It is a multi-letter cipher, developed by the mathematician Lester Hill in 1929.

For encryption, algorithm takes m successive plaintext letters and instead of that substitutes m cipher letters. In Hill cipher each character is assigned a numerical value
like: a=0,b=1, ... z=25.

The substitution of cipher text letters in place of plaintext leads to m linear equations. For m=3, the system can be described as follows:

$C1=(K_{11}P1+K_{12}P2+K_{13}P3)MOD26$---------
$C1=(K_{21}P1+K_{22}P2+K_{23}P3)MOD26$---------
$C1=(K_{31}P1+K_{32}P2+K_{33}P3)MOD26$---------

This can be expressed in terms of column vectors and matrices: C=KP Where C and P are column vectors of length 3, representing the plaintext and the cipher text and K is a 3*3 matrix, which is the encryption key. All operations are performed mod 26 here. Decryption requires the inverse of matrix K. The inverse K^{-1} of a matrix K is defined by the equation.

$K K^{-1}=$ I where I is the Identity matrix.

NOTE: The inverse of a matrix doesn't always exist, but when it does it satisfies the proceeding equation.

K^{-1} is applied to the cipher text, and then the plain text is recovered. In general terms we can write as follows: For encryption: C=P*K

For decryption: $P=C*K^{-1}$

Novel Modification to the Algorithm

As we have seen in Hill cipher decryption, it requires the inverse of a matrix. One problem arises that is:
Inverse of the matrix doesn't always exist. Then if the matrix is not invertible then encrypted text cannot be decrypted.
In order to overcome this problem author suggests the use of self-repetitive matrix. This matrix if multiplied with itself for a given mod value (i.e. mod value of the matrix is taken after every multiplication) will eventually result in an identity matrix after N multiplications. So, after N+ 1 multiplication the matrix will repeat itself. Hence, it derives its name i.e. self-repetitive matrix. It should be non-singular square matrix.

Modular Arithmetic: A brief Analysis

The analysis presented here for generation of self-repetitive matrix is valid for matrix of positive integers that are the residues of modulo arithmetic on a prime number. So, in analysis the arithmetic operations presented here are addition, subtraction, Unary operation, multiplication and division.

Existing System

Public-Key Cryptography

Public-Key Algorithms are symmetric, that is to say the key that is used to encrypt the message is different from the key used to decrypt the message. The encryption key, known as the Public key is used to encrypt a message, but the message can only be decoded by the person that has the decryption key, known as the private key.

This type of encryption has a number of advantages over traditional symmetric Ciphers. It means that the recipient can make their public key widely available- anyone wanting to send them a message uses the algorithm and the recipient's public key to do so. An eavesdropper may have both the algorithm and the public key but will still not be able to decrypt the message. Only the recipient, with the private key can decrypt the message.

An advantage of public-key algorithm is that they are more computationally intensive than symmetric algorithms, and therefore encryption and decryption take longer. This may not be significant for a short text message, but certainly is for bulk data encryption.

The Basic Principle

In order to decrypt a message, Bob (the recipient) has to know the key. However, it may be difficult for Alice (the sender) to tell Bob what the key is. If they simply agree on a key bye-mail for example, Eve could be listening in on their e-mail conversation and thus also learn what the key is. Public key cryptography was invented to solve this problem.

When using public-key cryptography, Alice and Bob both have their own key pairs. A key pair consists of a public key and a private-key. If the public-key is used to encrypt something, then it can be decrypted only using the private-key. And similarly, if the private-key is used to encrypt something, then it can be decrypted only using the public-key. It is not possible to figure out what the private-key is given only the public-key, or vice versa.

This makes it possible for Alice and Bob to simply send their public keys to one another, even if the channel they are using to do so is insecure. It is no problem that Eve now gets a copy of the public keys. If Alice wants to send a secret message to Bob, she encrypts the message using Bob's public key. Bob then takes his private key to decrypt the message. Since Eve does not have a copy of Bob's private key, she cannot decrypt the message. Of course, this means that Bob has to carefully guard his private key. With public key cryptography it is thus possible for two people who have never met to securely exchange messages. Figure 1 illustrates the public-key encryption process.

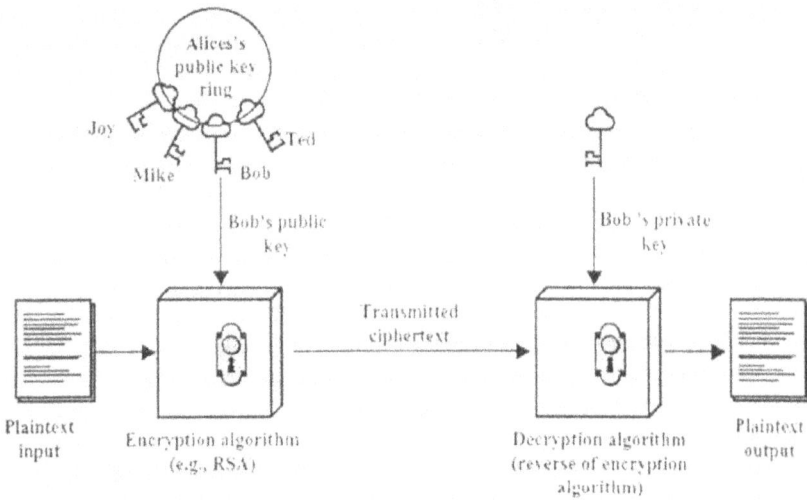

Figure 1. Encryption.

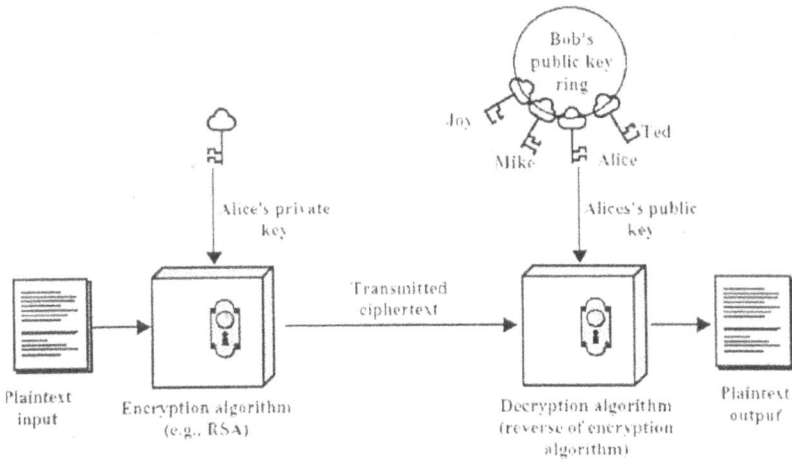

Figure 2. Public Key Encryption.

Advantages and Disadvantages of Public-Key Cryptography Compared with Secret-Key Cryptography

The primary advantage of public-key cryptography is increased security and convenience. Private keys never need to transmit or revealed to anyone. In a secret-key system, by contrast, the secret keys must be transmitted (either manually or through a communication channel), and there may be a chance that an enemy can discover the secret keys during their transmission.

Another major advantage of public-key systems is that they can provide a method for digital signatures. Authentication via secret-key systems requires the sharing of some secret and sometimes requires trust of a third party as well. As a result, a sender can repudiate a previously authenticated message by claiming that the shared secret was somehow compromised by one of the parties sharing the secret. For example, the Kerberos secret-key authentication system involves a central database that keeps copies of the secret keys of all users; an attack on the database would allow widespread forgery. Public-key authentication, on the other hand, prevents this type of repudiation; each user has sole responsibility for protecting his or her private key. This property of public-key authentication is often called non-repudiation.

A disadvantage of using public-key cryptography for encryption is speed; there are popular secret-key encryption methods that are significantly faster than any currently available public-key encryp-

169

tion method. Nevertheless, public-key cryptography can be used with secret-key cryptography to get the best of both worlds. For encryption, the best solution is to combine public- and secret-key systems in order to get both the security advantages of public-key systems and the speed advantages of secret-key systems. The public-key system can be used to encrypt a secret key, which is used to encrypt the bulk of a file or message. Such a protocol is called a digital envelope.

Public-key cryptography may be vulnerable to impersonation, however, even if users' private keys are not available. A successful attack on a certification authority will allow an adversary to impersonate whomever the adversary chooses to by using a public-key certificate from the compromised authority to bind a key of the adversary's choice to the name of another user.

In some situations, public-key cryptography is not necessary and secret-key cryptography alone is sufficient. This includes environments where secure secret - key agreement can take place, for example by users meeting in private. It also includes environments where a single authority knows and manages all the keys (e.g., a closed banking system) Since the authority knows everyone's keys already, there is not much advantage for some to be "public" and others "private" Also, public-key cryptography is usually not necessary in a single-user environment. For example, if you want to keep your personal files encrypted, you can do so with any secret-key encryption algorithm using, say, your personal password as the secret key. In general, public-key cryptography is best suited for an open multi-user environment.

Public-key cryptography is not meant to replace secret-key cryptography, but rather to supplement it, to make it more secure. The first use of public-key techniques was for secure key exchange in an otherwise secret-key system; this is still one of its primary functions. Secret-key cryptography remains extremely important and is the subject of ongoing study and research. Some secret-key cryptosystems are discussed in the sections on Block Cipher and StreamCipher.

ElGamal Public Key System

The ELGamal cryptographic algorithm is a public key system like the Diffie-Hellman system. It is mainly used to establish common keys and not to encrypt message. The ElGamal cryptographic algorithm is comparable to the Diffie-Hellman system. Although the inventor, Taher Elgamal, did not apply for a patent on his invention, the owners of the Diffie-Hellman patent felt this system was covered by their patent. For no apparent reason everyone calls this the IElGamal" system although Mr. ElGamal's last name does· not have a capital letter'G'[19]

A disadvantage of the EI Gamal system is that the encrypted message becomes very big, about twice the size of the original message m. For this reason, it is only used for small messages such as secret keys.

Generating the EI Gamal Public key

As with Diffie-Hellman, Alice and Bob have a (publicly known) prime number p and a generator g. Alice chooses a random number a and computes $A = ga$. Bob does the same and computes $B = g^b$. Alice's public key is A and her private key is a. Similarly, Bob's public key is B and his private key is b.

Encrypting and Decrypting Messages

If Bob now wants to send a message m to Alice, he randomly picks a number k, which is smaller than p. He then computes:

$C_1 = g^k \bmod p$

$C_2 = A^{k*} m \bmod p$

And send C_1 and C_2 to Alice. Alice can use this to reconstruct the message m by computing

$C_1 \qquad\qquad {}^{-a*} C_2 \bmod p = m$

Drawbacks

The existing system is very complex during encryption and decryp-

tion process.
Processing overhead of Encryption and decryption process is also more.

Proposed System

Elliptic Curve Crypto-system

Elliptic Curve Cryptography (ECC) is a suitable for key generation, encryption and decryption, because it can provide high security with smaller key size and lower power consumption.

An elliptic curve E over a prime field F_p is defined by
$y^2 \equiv x^3 + ax + b \pmod{p}$ Where $\quad$ a, b $\in F_p$, p $\neq$ 2,3, and satisfy the condition

$$4a^3 + 27b^2 \neq 0 \pmod{p}$$

The elliptic curve group $E(F_p)$ consists of all points (x,y) that satisfy the elliptic curve E.

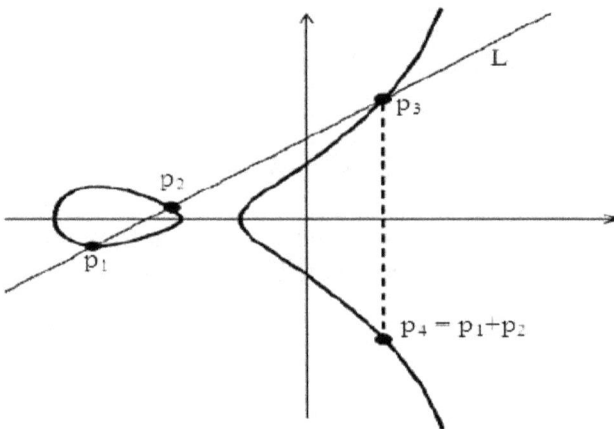

Figure 3. Elliptic curve cryptography.

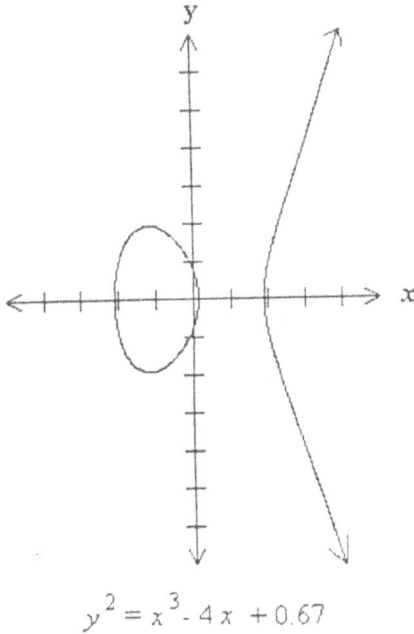

$$y^2 = x^3 - 4x + 0.67$$

Figure 4. Elliptic curve cryptography.

Elliptic Curve Operations

It consists of three operations:
Point Addition
Point Doubling
Scalar Multiplication

Point Addition

Suppose $P_1 = (x_1, y_1)$ and $P_2 = (x_2, y_2)$ where $P_1 \neq P_2$ are two points lie on an elliptic curve E. Adding the two points P_1 and P_2 giving a third point R that should lie on the same curve E.
$R = P_1 + P_2$

$s = (y_2 - y_1) / (x_2 - x_1)$

$x_3 = (s^2 - x_1 - x_2) \pmod p$

$Y_3 = (s(x_1-x_3)-y_1)$ (mod p)

Point Doubling

Adding the pointP = (x_1,y_1) that lies on the elliptic curve E to itself is called point doubling. The point R that results from doubling the point P is also lies on the elliptic curve E.

Where

$R = 2P = P + P =$ \qquad (x_3,y_3)

$s = (3x_1^2 + a) / (2y_1)$ $x_3 = (s^2 - 2x_1)$ (modp)
$Y_3 = (s(x_1-x_3)-y_1)$ (mod p)

Scalar Multiplication

The scalar multiplication of an integer k by the point Q = (x_1,y_1) that lies on the curve E can be defined by repeating the addition of the point Q to itself k times. The result point R also lies on the elliptic curveE.
R= kQ = Q+Q+Q+.....+Q (k times)
For example, 15Q can be done using point addition and point doubling.
15Q = 2 (2 (2Q + Q) + Q) + Q.

Hill Cipher

Hill cipher is invented by Lester S. Hill in 1929 and it is a polygraphic substitution cipher based on linear algebra. Each pixel is represented by a number modulo 256. Often the simple scheme the grey level or grey value indicates the brightness of a pixel. The minimum grey level is 0. The maximum grey level 255 is used, but this is an essential feature of the cipher. To encrypt a message, each block of n letters (considered as an n-component vector) is multiplied by an self-invertible n × n matrix, against modulus 256. To decrypt the message, each block is multiplied by the inverse of the matrix used for encryption. The matrix used for encryption is the cipher key, and it should be chosen randomly from the set of invertible n × n matrices (modulo256).

Where

C = Ciphertext

P = Plain text

K = Self-invertible key

[C] = [P]*[K]mod(256) (for encryption)

[P] = [C]*[K^{-1}]mod (256) (for decryption)

$$[P] = \begin{bmatrix} P1 \\ P2 \end{bmatrix}, \qquad \begin{bmatrix} C1 \\ C2 \end{bmatrix}, [K] \in \begin{bmatrix} k11 & k12 \\ \overline{k}21 & k22 \end{bmatrix}$$

To decrypt the ciphertext message C, the recipient needs to compute the key matrix inverse (K^{-1}) Where K*K^{-1}= I Iis the identity matrix, then use the following equation to produce the plaintext P
P= K^{-1} C mod 256

Proposed System Algorithm

The Proposed Approach (ECCHC)

Step 1: Key Generation

User A (The sender)

Choose the private key $n_a \in [1,e-1]$

Compute the public key $P_a = n_a.G$

Compute the initial key $K_1 = n_a.P_b = (x,y)$

Compute $K_1 = x.G = (k_{11},k_{12})$ and $K_2 = y .G = (k_{21},k_{22})$

Generate the self-invertible key matrix K_m

User B (The receiver)

Choose the private key $n_b \in [1, e-1]$

Compute the public key $P_b = n_b.G$

Compute the initial key $K_1 = n_b . P_a = (x,y)$

Compute $K_1 = x.G = (k_{11}, k_{12})$ and $K_2 = y .G = (k_{21}, k_{22})$

Generate the self-invertible key matrix K_m

Step 2: Encryption (User A)

Separate the original image pixel values into blocks of size four.

Arrange each block into four rows column vector $(4 \square 1)$.

Multiply the self-invertible key matrix K_n by each vector $(P_1, P_2, P_3 ...)$ and take modulo 256 for each value $C_1 = (K_m . P_1)$ mod 256.

Construct the ciphered image C from the values in the ciphered vectors $(C_1, C_2, C_3 ...)$.

Step 3: Decryption (User B)

Separate the ciphered image pixel values into blocks of size four.

Arrange each block into four rows column vector $(4 \square 1)$.

Multiply the self-invertible key matrix K_n by each vector $(C_1, C_2, C_3 ...)$ andtakemodulo256foreachvalue$P_1 = (K_m.C_1)$mod 256.

Construct the original image from the values in the deciphered vectors $(P_1, P_2, P_3 ...)$.

Software Implementation

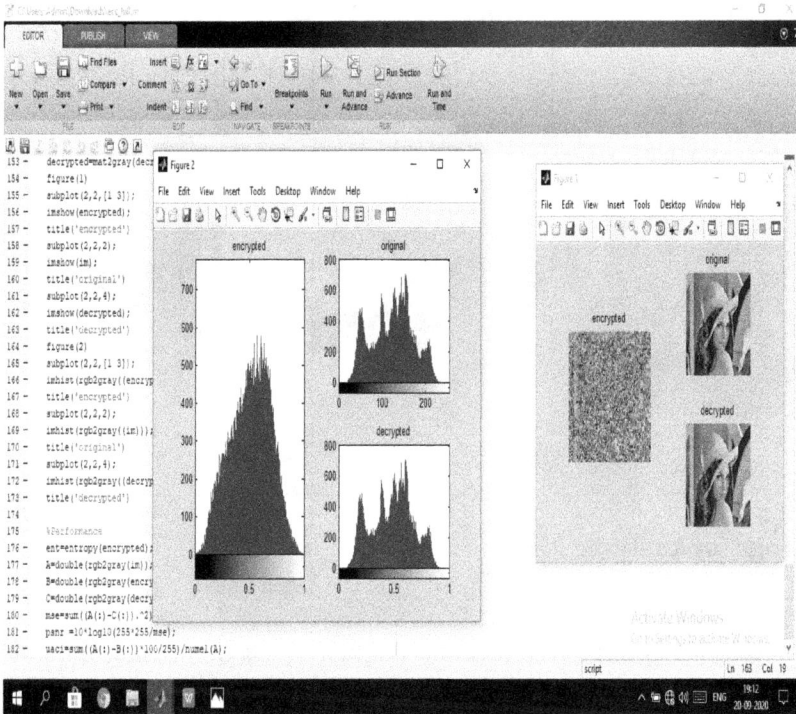

Figure 5. Screenshot of output.

Performance Analysis

This system is implemented in MATLAB 2014a and the results were analyzed and compared with various schemes to demonstrate the efficiency of our scheme. Our results show that the proposed scheme is more efficient than other schemes.

Table 1. Examples of encrypted and decrypted images

ORGINAL IMAGE	ENCRYPTED IMAGE	DECRYPTED IMAGE
original	encrypted	decrypted

Histogram Analysis

In image encryption environment, the histogram of an image generally states to a histogram of the pixel intensity values. This histogram is a graph display the quantity of pixels in an image at each dissimilar intensity value found in that image. For an 8-bit gray scale image there are 256 different possible intensities, and so the histogram will graphically display 256 numbers showing the distribution of pixels amongst those gray scale values. Thus, by relating the histograms of original and encrypted images, we can say that the encrypted images are uncertainty identical.

Table 2. Histogram analysis

Histogram of Lena image before Encryption	Histogram of Lena image after Encryption	Histogram of Lena image after Decryption

Entropy Analysis Table

Information entropy is the quantity used for uncertainty of an image. It can serve a measure of disorder. In place of the level of disorder increases, the entropy growths and system become less predictable. The entropy is given as,

$$E = \sum_{i=0}^{n} P(x) \times \log_2 P(x)$$

Where, X represents the test image, x_i symbolizes the i_{th} possible

179

value in X. Entropy of encrypted images using our algorithm and other references are shown in Table1.

Table 3. Entropy values for scheme and reference

Image Name	Our approach	Ref 3	Ref 4	Ref 5	Ref 6	Expect ed Value
Lena	7.9957	7.9961	7.9972	7.9976	7.9891	8

Theoretical entropy value for a gray scale image is 8 and it is expected to be near to 8. The entropy value calculated for our algorithm is 7.9963.

UACI

The number of pixels change ratio and unified average changing intensity (UACI) are the most important parameters used to evaluate the strength of image cryptosystems. A high UACI value means, the cryptosystem has high resistance to differential attacks.

Theoretically, UACI values are calculated using the following formulas.

$$UACI = \frac{1}{H \times G} \sum_{i=1}^{H} \sum_{i=1}^{G} \frac{(C1_{i,j} - C2_{i,j})}{R} \times 100\%$$

Where, $H \times G$ is the dimension of the images and it should be same for both original and encrypted images
$C1_{i,j}$ is the pixel value at i,j of the original image
$C2_{i,j}$ is the pixel value at i,j of the encrypted image

R is the maximum possible pixel value of the image dimension. In the proposed scheme, we approach a new scheme by establishing a mathematical model for image encryption and calculated NPCR and UACI values for this model with Lena image. Further, these values are used to analyze the strength of the cryptosystem. Experimental results for UACI tests from table2 and table3 show that the proposed cryptosystem is efficient and more secure with some other existing systems.

Table 4. UACI of proposed scheme and reference

Image name	Our approach	Ref 3	Ref 4	Ref 5	Ref 6	Expected Value
Lena	48.44	38.33	33.28	33.26	33.42	EXPECTEDTO HIGH

Mean Square Error

Mean Square Error (MSE) is a parameter used to measure difference between original and encrypted image in which pixels are expressed between 0 and 255. To check the quality of encryption the original image and cipher image is usually compared. MSE for original and cipher image can be calculated as:

$$\text{MSE} = \frac{1}{H \times G} \sum_{i=1}^{H} \sum_{j=1}^{G} (C1_{i,j} - C2_{i,j})^2$$

Where, $m \times m$ is the dimension of the image

$C1_{i,j}$ is the pixel values of original image

$C2_{i,j}$ is the pixel values of encrypted image

High value of MSE shows that the encrypted image has more variations of pixel values than original image.

Table 5. MSE of proposed scheme and reference

Image name	Our ap-proach	Ref 3	Ref 4	Ref 5	Expected Value
Lena	17537	111.22	NA	NA	EXPECTEDTO HIGH

PSNR

The PSNR is stands for Peak Signal to Noise Ratio. This ratio is frequently used as a quality and quantity among the original and an encrypted image. PSNR characterizes a quantity of the peak error. It is mathematically calculated as:

$$PSNR = 10 \log_{10} \frac{(R)^2}{MSE}$$

Where, R is the maximum possible value of the pixel
MSE is the Mean Squared Error between the original and encrypted image.

Table 6. PSNR of proposed scheme and reference

Image name	Our ap-proach	Ref 3	Ref 4	Ref 5	Expected Value
Lena	5.6912	27.6689	NA	NA	EXPECTED TO BE LOW

The higher value of MSE and lower value of PSNR means, the encryption algorithm is better for encryption and the randomness of encryption image pixel values are high.

Structural Similarity Index

The Structural Similarity (SSIM) index is a method for measuring the comparison between two images. The SSIM index can be observed as a quality measure of one of the images being compared with other image which is observed to ensure the similarity of original and decrypted image.

$$SSIM(x, y) = \frac{(2\mu_x\mu_y + c_1)(2\mu_{xy} + c_2)}{(\mu_{x^2} + \mu_{y^2} + c_1)(\mu_{x^2} + \mu_{y^2} + c_2)}$$

Where, μ_x and μ_y are mean of x and y.
σ_x, σ_y, and σ_{xy} are covariance of x, y, and xy
c_1 and c_2 are constants.
Our proposed system achieved SSIM value 1 and it shows that,
SSIM value is calculated by comparing the original image and decrypted image and it is expected to be 1. Our result shows that the image is decrypted using our proposed encryption algorithm without error.

Conclusion

In this proposed scheme, the encryption algorithm is implemented with modified hill cipher approach and one reference images Lena were tested using MATLAB2014a. Various performance parameters were calculated and analyzed with various schemes. Our entropy value 7.9957for the cipher image shows that our proposed image encryption algorithm is efficient. The calculated SSIM value of 1 show that the original image and decrypted image are similar and there is no error while encryption and decryption process. Other performance parameters such as MSE, PSNR, and UACI calculated for our proposed algorithm show that, our new proposed scheme is efficient and more secure for image encryption process.

References

[1]   Ziad E. Dawahdeh, Shahrul N. Yaakob, Rozmie Razif bin Othman, A new image encryption technique combining Elliptic Curve Cryptosystem with Hill Cipher, Journal of King Saud University - Computer and Information Sciences, (2017). https://doi.org/10.1016/j.jksuci.2017.06.004.

[2]   Hossam Diab, Aly M. Elsemary. Secure Image Cryptosystem with Unique Key Streams viaHyper-chaotic System, Signal Processing, (2017). doi:10.1016/j.sigpro.2017.06.028.

[3]   Naveen Kumar S K, Panduranga H T. Advanced Partial Image Encryption using Two-Stage Hill Cipher TechniqueInternational Journal of Computer Applications (0975 – 8887) Volume 60– No.16, December 2012

[4]   Wang, Xingyuan & Wang, Qian & Zhang, Ying-Qian. (2014). A fast image algorithm based on rows and columns switch. Nonlinear Dynamics. 79. 1141-1149. 10.1007/s11071-014-1729-y.

[5]   Wenhao Liu, Kehui Sun, Congxu Zhu. A fast image encryption algorithmbasedon chaotic map. OpticsandLasersinEngineering10.1016/J.OPTLASENG.2016.03.019

[6]   Xiao Chen, Chun-Jie Hu. Adaptive medical image encryption algorithm based on multiple chaotic (2017)https://doi.org/10.1016/j.sjbs.2017.11.023

[7]   Congxu Zhu. A novel image encryption scheme based on improved hyper-chaotic sequences, Optics Communications 285 (1) (2012) 29--37.doi:10.1016/j.optcom.2011.08.079.

[8]   Xiao, Chun-Jie Hu, Adaptive medical image encryption algorithm based on multiple chaoticMapping, Saudi Journal of Biological Sciences 24 (2017)1821–1827.

[9]   Deng, S.J., Huang, G.C., Chen, Z.J., 2011. Research and implement of Self adaptive image encryption algorithm based on chaos. J. Comput. Appl. 31 (6),1502–1504.

[10]   Zhang, J., Hou, D., Ren, H., 2016. Image Encryption Algorithm Based on Dynamic DNA Coding and Chen's Hyperchaotic System, Mathematical Problems in Engineering, Article ID 6408741,11pages.

[11]   Zhongyun Hua, Fan Jin, Binxuan Xu, Hejiao Huang, 2D Logistic-Sine-Coupling Map for Image Encryption, Signal Processing

(2018), doi:10.1016/j.sigpro.2018.03.010.

[12] Hui Wang, Di Xiao, Xin Chen, Hongyu Huang, Cryptanalysis and Enhancements of Image Encryption Using Combination of the 1D Chaotic Map, Signal Processing (2017), doi: 10.1016/j.sigpro.2017.11.005.

[13] Rushi Lan, Jinwen He, Shouhua Wang, Tianlong Gu, Xiaonan Luo, Integrated Chaotic Systems for Image Encryption, Signal Processing (2018), doi:10.1016/j.sigpro.2018.01.026.

[14] SakshiDhall , Saibal K. Pal , Kapil Sharma , Cryptanalysis of image encryption scheme based on a new 1D chaotic system, Signal Processing (2017), doi:10.1016/j.sigpro.2017.12.021.

[15] Zhongyun Hua, Shuang Yi, Yicong Zhou, Medical image encryption using high-speed scrambling and pixel adaptive diffusion, Signal Processing (2017), doi:10.1016/j.sigpro.2017.10.004.